On Sex and Gender

A Commonsense Approach

Doriane Lambelet Coleman

Simon & Schuster

New York   Toronto   London   Sydney   New Delhi

100 YEARS
SIMON &
SCHUSTER

1230 Avenue of the Americas
New York, NY 10020

First Simon & Schuster hardcover edition May 2024

SIMON & SCHUSTER and colophon are registered trademarks of Simon & Schuster, LLC

Simon & Schuster: Celebrating 100 Years of Publishing in 2024

For information about special discounts for bulk purchases,
please contact Simon & Schuster Special Sales
at 1-866-506-1949 or business@simonandschuster.com.

The Simon & Schuster Speakers Bureau can bring authors to your live event.
For more information or to book an event, contact the
Simon & Schuster Speakers Bureau at 1-866-248-3049
or visit our website at www.simonspeakers.com.

Interior design by Ruth Lee-Mui

Manufactured in the United States of America

1   3   5   7   9   10   8   6   4   2

Library of Congress Cataloging-in-Publication Data is available.

ISBN 978-1-6680-2310-5
ISBN 978-1-6680-2312-9 (ebook)

For Bluette, who taught me to have
an unbounded self-conception

Chama had to wait months to get the exact red silk she was looking for, and then the matching blue a few weeks later, and even then the colors were not quite right. She and Sidi Allal did not mean the same thing by "red" and "blue." People, I discovered, often did not mean the same thing by the same word, even when talking about seemingly banal things like colors.

—Fatima Mernissi,
Dreams of Trespass:
Tales of a Harem Girlhood

Contents

Introduction   xi

Part I: What Is Sex?   1
1. The Answer from Biology   3
2. The Answer from Law   34
3. The Answer from Progressive Advocacy   55

Part II: Sex Matters   83
4. Sex Is Good!   85
5. Sex Just Is (Like Age)   114
6. Sex Is Still a Problem (Like Race)   136

Part III: On Sex and Gender   165
7. The (Un)Lawfulness of Regulating on
   the Basis of Sex and Gender   167
8. The Politics of Sex and Gender   198
9. A Commonsense Approach   232

Acknowledgments   265
Bibliography   269
Notes   271
Index   295

Introduction

ON APRIL 2, 2019, the Judiciary Committee of the United States House of Representatives held a hearing to consider legislation called the Equality Act. The act, which goes by the bill number H.R. 5 in sessions when the Democrats hold the House, is an initiative designed to secure federal civil rights protections for people who are gay or transgender. It would ensure, for example, that a pediatric practice couldn't refuse to enroll a baby as a patient because he has two moms, and a hotel couldn't refuse to accept a booking from someone because they're transgender. Many states already have these protections, but many don't, and discrimination based on sexual orientation and gender identity is ongoing. A federal statute would provide national-level assurances that people in the LGBTQ communities have the right—like everyone else—to be treated with dignity and respect as they go about their lives.

These protections are no-brainers as far as I'm concerned. A person's gender identity and sexual orientation have nothing to do with whether they should be able to buy bread (a necessary transaction) or rent skis (a discretionary transaction). Basic decency isn't a partisan proposition, and tolerance of individual difference is a minimal requirement for a well-functioning society—especially a pluralistic one like the United States. Whether I'm a buyer or a seller, your humanity—not your politics—tells me that I should engage with you courteously. As children, we're taught this as the golden rule: treat others as you would want them to treat you. As adults, the lesson is no less golden, and everyone depends on it every day.

Nonetheless, I wasn't in Washington to speak in favor of the Equality Act. I'd been invited to the hearing by Republicans who oppose H.R. 5. They wanted me to testify to its implications as drafted for girls' and women's sports. These were at risk because the strategy adopted by the bill's sponsors to secure rights for gay and trans people wasn't simply to prohibit discrimination based on sexual orientation and gender identity. Instead, they had included two radical proposals that would go far beyond what was necessary to achieve that goal. I'm not using the word *radical* politically here, I'm using it literally to mean—per the Oxford Dictionaries via Google—"affecting the fundamental nature" of something.

The first of these proposals was to re-define *sex* in federal law in a way that bears little resemblance to how the word is normally used. By *sex* we usually mean—per Oxford again—"either of the two main categories (male and female) into which humans and most other living things are divided on the basis of their reproductive functions." Instead, the drafters of H.R. 5 defined it as "a sex stereotype; pregnancy, childbirth, or a related medical condition; sexual orientation or gender identity; and sex characteristics, including intersex traits." This was essentially the wish list from progressive advocates pushing for a sex-blind society— one in which sex is deconstructed and certain core truths are discarded. Whatever your political leanings, it would undoubtedly affect the fundamental nature of things as we know them to disconnect sex from the male and female body and also from reproduction more generally.

The second proposal was to prohibit—without exception—all sex classifications. Even good and valuable ones. Again, whatever your political leanings, going sex-blind in law and policy would undoubtedly affect the fundamental nature of things. Sex is our first natural taxonomy, meaning the first way humans are divided biologically. It's probably also our first social taxonomy, that is, the first way we sort people and are sorted by others. Lots of traditional sex classifications have been properly discarded because they were wrong or simply designed to subordinate the female half of the population. But you don't have to be a traditionalist or conservative—just human—to know that not all decisions based on sex are sexist.

Sports is an obvious example. If H.R. 5 had passed as drafted, it wasn't clear that we could continue to have "women's" and "men's" sports—at all. Even if we could, though, it's clear that sorting people into and out of these groups by sex would be prohibited by the legislation. In that case, it would defy common sense for a policymaker to support two parallel programs, both of which include a mix of males and females. For some, this is precisely the point: as with the push for gender-neutral restrooms, their utopia is unisex.

I was on the list of witnesses to testify not just because I had decades of experience with sex-segregated sport, but because I'd taken public stances in those years about the value of protecting the female category.

I'd been an elite athlete. My freshman year in college—at Villanova— was 1978, the first year American universities were required by Title IX to begin supporting women's teams and awarding sports scholarships also to female athletes. Had these mandates not been set, only men would have had the physically, educationally, and culturally empowering experiences that shaped me and my peers. In 1982, my senior year in college—at Cornell—I became one of the first student-athletes to graduate from the Ivy League as a collegiate national champion, with the now-defunct Association for Intercollegiate Athletics for Women. That association ran women's sports nationwide until the NCAA deigned to add us to their programming. For several years thereafter, I competed for my club and national teams.

As a young lawyer starting in the early 1990s, I helped to develop anti-doping programs and to prosecute cases brought under those programs. It's because I was an elite athlete and have been working with scientific experts in androgens for decades that I know a lot about their dual body-building and performance-enhancing effects. As part of this ongoing work, in 1999, I testified before the Senate Commerce Committee at the request of its chair, John McCain, in connection with the establishment of the World Anti-Doping Agency (WADA). I then worked with the Clinton White House on its contribution to WADA's founding documents.

As a legal academic, I've written and advised regulators about how empowering women and girls through sports has done enormous individual and societal good that can be realized only when athletes are sorted by sex. In 2017 specifically, I published a law review article called "Sex in Sport," which focuses on the challenge to the sex-based eligibility rules for the female category. This article was the basis for my testimony earlier in 2019 in Caster Semenya's case at the Court of Arbitration for Sport, which I discuss in chapter 1.

I'm a liberal Democrat. Back in 1999, it never occurred to me to be concerned about testifying for "the other side" because the anti-doping effort wasn't a partisan proposition. In those days, no one who wasn't cheating wanted excess androgens in girls' and women's sports. Democrats and Republicans alike understood that sex-linked biology is central to success in competitive sports, and they were all in for supporting female athletes.

In 2019, the issue was a version of the exact same question, but everything was suddenly fraught. I was concerned about testifying for the Republicans because their goals then were very different than mine. It's an understatement to say that rights for gay and transgender people aren't a bipartisan commitment. Leaving aside the desire for a sex-blind society that motivates H.R. 5's most radical backers, the day-to-day obstacles for people in the LGBTQ communities are real and, as I would say at the top of my remarks before the House Judiciary Committee, I support equality for everyone. Many Republicans support equality too, of course, but not those who were calling for my testimony.

I was further concerned because by testifying I would be breaking ranks with my people on an important political question, and I would be exacerbating a fault line that opponents of things that matter to me could exploit. The bill's sponsors and backers were undeniably my people, and my commitment to equality goes deeper than the levers I pull on Election Day. I'm mixed-race, I was raised in a civil rights family, and my husband, who is African American, is a civil rights lawyer with decades-old ties to the social justice movements.

The problem was that no one from "my side" was giving women who reject sex blindness the time of day. In April 2019, it seemed they were either locked into sex blindness as their approach (to trans rights) or it was their goal (for society more generally), and they were brooking no dissent. Deplatforming, censorship, and cancellation of women like me were rife.

My sense, though, was—and remains—that most people in the broader Democratic family, along with many moderate Republicans, want civil rights protections for gay and trans people, but not by way of denying sex. Outside of certain academic and political bubbles, people know that sex is real; they know that it matters to their lives; and—importantly—they know that it's not only about discrimination and equality so that filtering it exclusively through that lens gives you a warped sense of things. My sense is that most people aren't interested in a sex-blind society; they're interested in a sex-smart society. My hope was that if I accepted the GOP's invitation to testify, I could speak for a nonideological, commonsense, and evidence-based policy.

And so, on a cool spring day in Washington, D.C., in a packed room full of members of Congress, witnesses, legislative aides, and the press, I began my testimony with a general point about *inequality*:

[D]ifferent groups experience inequality for different reasons, at the hands of different people, and in different ways, so that tailoring an effective remedy requires careful attention to those differences. Although the nation benefits as equality expands, in fact only some amongst us needed the Emancipation Proclamation and *Brown v. Board of Education*. Only some of us need Title IX and the Violence against Women Act. Approaches to addressing equality that elide relevant differences are not only ineffective; they can actually serve as cover for ongoing inequality.

I then focused on the problem with the way H.R. 5 was drafted:

I support equality, including for the LGBTQ community. But I don't support the current version of H.R. 5 because—and I say this with enormous respect for everyone who cares about and is working on the bill—it elides sex, sexual orientation, and gender identity: It's all sex discrimination, and, at least impliedly, we're all the same. In opting for what is in effect a sex-blind approach to sex discrimination law, the legislation would serve as cover for disparities on the basis of sex.

I explained that sport provides an easy example of how sex-blind law would disparately affect females, and how we sometimes have to act on the basis of sex to achieve equality:

Scientists agree that males and females are materially different with respect to the main physical attributes that contribute to athletic performance, and they agree that the primary reason for sex differences in these attributes is exposure in gonadal males to much higher levels of testosterone during growth and development (puberty), and throughout the athletic career.

This different exposure literally builds the male body in the respects that matter for sport.

If U.S. law changes so that we can no longer distinguish between females and women with testes for any purpose, we risk not knowing the next Sanya Richards-Ross or the next Allyson Felix. We risk losing the extraordinary value that comes from having Serena Williams, Aly Raisman, and Ibtihaj Muhammad in our lives and on the medal stand. If they bothered to compete, they would be relegated to participants in the game.

To the argument that participation should be enough—that females don't also need to win—I added:

Participation contributes to equality, but the real power of girls and women in sport isn't in gym class, it's in teams, in competitions, and in victories. It's in the same numbers of scholarships and spots in finals and on podiums. It's in the fact that Brandi Chastain can win World's, celebrate like the guys, and get a whole generation of little girls to play soccer because she did. It's in the fact that Simone Manuel can win Olympic Gold in the 100 meters freestyle with millions watching on prime-time television and from there can lead a generation of African American kids to the pool who didn't believe that swimming was for them too.

I concluded my remarks by encouraging members of Congress "to consider revisions to H.R. 5 that provide protections for LGBTQ people that don't risk these invaluable goods, and that are otherwise considered about the circumstances in which sex still matters."

Taking the counter position that day was the legal director of the National Women's Law Center (NWLC), and her remarks—at least for me personally—were brutal. The NWLC has long been one of America's leading organizations protecting women's rights. If any group knew that the discrimination and disparities women face have always been based in the female body—and that ignoring this has always served to perpetuate our inequality—it was the NWLC. It's not an understatement to say that for decades the organization was my champion. (Also, for a time in the 1980s, my husband was on its board.) I knew going into the hearing that a few years earlier, it had started taking trans-inclusive positions—but so had I, and I had never found it necessary to deny the facts of the female body, the science of sex differences, or common sense and experience. But at the hearing, its legal director did just that.

She was unequivocal that "[t]rans women are women, period." By "period" she meant three things: (1) no discussion, (2) for all purposes, and (3) no connection to female biology required. At the time, the first two were emerging refrains, and progressive women's organizations had already started shutting out those of us who wanted to be inclusive but also to talk about evidence-based exceptions. But the third—no connection to female biology required—was new, at least outside of academia and radical advocacy groups. It's difficult to overstate its historical significance.

The Woman Question has a centuries-long pedigree. Before the trans rights movement got a hold of it, it had always been about separating what was true about the female sex from the social constructions that were designed to keep us down. *Those* were to be discarded, not the truth. Inside certain academic and advocacy circles you'd hear about "the different journeys to womanhood" but on the outside we hadn't divorced what it means to be a woman from female biology. Indeed, to that point, for most people, transgender women and girls were "transgender"—and "women" if you were comfortable with that, as I was—precisely because of their physical transitions. It was revolutionary for a mainstream women's rights organization to take an eraser to female biology as the linchpin to womanhood.

The NWLC's legal director also repeatedly elided the difference between sex and gender, describing real sex differences as "just fears and myths and stereotypes" and invoking feminist icon Ruth Bader Ginsburg's words and brand as she claimed this "is not a way that we can make law" and "we do not create policy about myths and stereotypes." Ginsburg didn't deny the reality of sex; she was all about distinguishing the facts from artifice. She didn't call for sex blindness; to the contrary, her most famous opinion on the topic—*United States v. Virginia* (1996)—stands for dismantling structural sexism while expressly leaving room for the law to be sex-smart, including to celebrate sex as well as to use it to promote sex equality and to empower the citizenry. Denying facts and science is never a good idea whether that denial comes from the right or the left.

In the legal director's concluding remarks, she invoked her organization's influential brand to urge people who might have been confused about what was going on—and it was hard not to be—to trust the positions it was taking on H.R. 5 because, she said, we are "the experts" on what's "good for women." Here's the quote in context:

> This is what we do, day in and day out, across sectors, workplace, you know, healthcare. Workplace, justice, education, all of these areas, this is what we do is we fight for women's rights. And so please look to us as the experts on whether or not this bill is good for women and LGBTQ people.

It was stunning not least in its apparent hubris and misogyny.

I never got the chance to respond that day, and H.R. 5 didn't get through Congress, but the culture war between those on the left who want to erase sex and those on the right who want to erase gender diversity has only ratcheted up. As a candidate for the 2020 Democratic presidential nomination, Joe Biden announced that the Equality Act would be his top legislative priority, and he's followed through to the extent of his authority. Florida governor

Ron DeSantis, who was a candidate for the 2024 Republican nomination, illustrates the response. He's spearheaded laws that ban gender-affirming care for trans kids and any instruction on gay and transgender issues in schools. It's not just straight male politicians who are presenting us with either/ors. In the sports world, Megan Rapinoe "would 'absolutely' welcome a transgender woman onto the USWNT," while Caitlyn Jenner "opposes biological boys who are trans competing in girls' sports in school." These are just a few prominent examples. But in between, where the polling shows you'll find most Americans, there's confusion and concern.

I'm writing this book for everyone who wants to understand what's going on for themselves, and who's inclined to be both inclusive and true to science and common experience. I'm also writing to say my piece. We are at a crossroads in the history of sex and gender. We've overcome a lot of the historical sexism that defined women and their lives, and we now need to decide if, going forward, we're going to be sex-blind or sex-smart. This was already the question in 1996 when Ginsburg penned the majority opinion in *United States v. Virginia*; that is, it predates the current trans movement. But in pushing as hard as it does for sex blindness, the movement has forced women—and men—in the middle to articulate a nonreligious case that sex matters, and to show that there's a path forward that embraces both sex and gender. These are the goals of this book.

Three important notes before you dig in, about context, language, and the ever-shifting ground.

On *context*: I made my life in the United States, and the American experience with sex and gender is at the center of this book. But I'm also a dual national, and I work a lot with people abroad. This is because the issues I focus on—sex discrimination law, elite sports, and scientific research—aren't constrained by national borders. Neither are the politics of sex and gender. On the one hand, trans-inclusive feminists have brought their cause into the world of international human rights; on the other, regressive authoritarian regimes have cracked down on LGBTQ rights and described their crusade as a fight for civilization. The fine points in any given country are undoubtedly different, but like sex and gender themselves, the themes are ubiquitous, as are the political arguments. I've worked to make this book accessible wherever you live.

On language: If you're reading this, you already know that the words we use to talk about sex and gender are contested, starting with *sex* and *gender* themselves—but also *woman* and *female*, *man* and *male*, *transgender*, and so on. Because language—the words we have and how we define them—affects what we can communicate and what we understand, the people who run movements understandably want to control it. The left is much more aggressive and organized about this, but the effort is made on both sides. In this book, I try to define the words as I go and explain how I'm using them, but in general, my goal is neither to be disrespectful of nor to pander to one side or the other. Rather, it's to speak freely and honestly; to communicate, not to obfuscate; and to reach people who want to learn and engage wherever they are.

On shifting ground: As I was writing this book, there wasn't a day that passed without a news article, book, or policy development that affected what I would say or how I would say it. Among the most important events were two Supreme Court decisions interpreting federal law: the first protecting gay and trans people from employment discrimination, the second abolishing the federal constitutional right to abortion. But there were also challenges to the medical standard of care for treating trans kids; to the ways teachers can talk about sex and gender in schools; and to the ways words like *sex* and *gender* are used in legacy media. Even the way information about sex and gender is collected and made available changed with artificial intelligence (AI). ChatGPT stopped me for a day as I looked up words and realized that no matter how hard we try, communication in this space—which is already difficult—will become even more elusive going forward. The ground shifts, but as you'll see in my first chapter, the basic facts are universal and timeless.

Part I

What Is Sex?

1

The Answer from Biology

CASTER SEMENYA OF SOUTH AFRICA is a two-time Olympic gold medalist and a three-time world champion at 800 meters. For a decade, from 2009 to 2019, she dominated this grueling two-lap race on the track in global competition. When she was healthy, she was unbeatable. As sport scientist Ross Tucker put it in 2016, "There is no more certain gold medal in the Rio Olympics than Semenya. She could trip and fall, anywhere in the first lap, lose 20 [meters], and still win the race." Indeed, a headline afterward read "Caster Semenya destroys rest of field to claim easy gold in women's 800m final." A few years later, she herself said simply, "I am the greatest that has ever done it."

No matter where you live, you've likely heard of Semenya. An Internet search brings up an astonishing amount of coverage in the world's newspapers, and she's been a particular favorite on Twitter (now X) and in academia. But the coverage is rarely about her sporting achievements or her remarkable dignity.

I come out of the world of track and field—*athletics* as it's known outside of the United States—and the 800 meters was also my best event. I wasn't a world or Olympic champion like Semenya, but I won three national championships and competed at the international level for several years. In 1983, I was in the race when Jarmila Kratochvilová of Czechoslovakia broke the still-standing world record. I'd love to be able to say that Kratochvilová and our other stars garner the attention Semenya receives, but we're just not that popular.

Semenya is famous because she's the global face of the cultural, political, and legal battle over the questions that are at the center of this book: What is sex? Is it about our male and female forms—including our reproductive biology—or is it now about something else? Is it still acceptable for governments and policymakers to use sex in this sense as a basis for regulation, or has biology ceased to matter any differently from, say, our height or our hair color? Semenya has come to represent these questions and the surrounding debates because, though her birth certificate reads "female" and she describes herself as a woman, she has, from the time she was a child, been perceived by others as male.

In a long profile of Semenya in the *New Yorker* in 2009, Ariel Levy shows that this was true of those in her village with whom she played football (soccer) as a child, as well as those against whom she competed on the track as a young teenager. As the trainer at her primary school explained to Levy:

> "[W]herever we go, whenever she made her first appearance, people were somehow gossiping, saying, 'No, no, she is not a girl,'" Phineas Sako said, rubbing the gray stubble on his chin. "'It looks like a boy'—that's the right words—they used to say, 'It looks like a boy.' Some even asked me as a coach, and I would confirm: it's a girl. At times, she'd get upset. But, eventually, she was just used to such things." Semenya became accustomed to visiting the bathroom with

4

a member of a competing team so that they could look at her private parts and then get on with the race. "They are doubting me," she would explain to her coaches, as she headed off the field toward the lavatory.

A 2009 article in the Guardian detailed that "Semenya's grandmother, Maphuti Sekgala, said Caster had been teased about her masculine appearance since the day she joined the village football team as the only girl" and that Michael Seme, her coach, was "well used to the commotion. He recalls stopping to use the facilities at a petrol station in Cape Town" and "as Semenya tried to enter the women's toilets, she was stopped by the petrol attendants." As an adult, she has been described by black African people in African publications as having a "masculine phenotype" and "man-like physical features."

As Semenya herself tells the story, the first time she met Violet Raseboya, the black woman who is now her wife, "She thought I was a boy": "We met in a restroom in 2007. She was a runner and was being escorted by doping officials." She said, 'What is a boy doing in here?'"

Despite what you may have heard, to code Semenya as male isn't racist or ignorant, it's the normal reaction to her face, voice, and body; to the way she holds herself; and to the way she moves. In competition, it's also the reaction to her seemingly effortless power over the last 200 meters of a race. The video of her victory at the 2009 World Championships in Berlin is probably the best evidence of that different power. It was her first global competition and she was just eighteen. As she comes off the turn and into the final straightaway, she seems startled to have so easily left her competitors—the very best females on the planet—in the dust. In the end, she gapped the field by more than 20 meters.

Semenya has always chosen to dress and behave in a masculine way, and this reinforces the sense of her physical presence. For example, her primary school trainer, Phineas Sako—whom Levy described as a "fluent but rough" speaker of English—seems to have coded her as male in part for these reasons. "Caster was very free when he is in the male company," Sako said. "I remember one day I asked her, 'Why are you always in the company of men?' He said, 'No, man, I don't have something to say to girls, they talks nonsense. They are always out of order.'" The headmaster at her high school seems to have done the same: "She was always rough and played with the boys. She liked soccer and she wore pants to school. She never wore a dress. It was only in Grade 11 that I realized she is a girl." When Semenya crossed the finish line at World's in 2009, she flexed her arms for the photographers, and her masculine "mannerisms" along with her "victory" and her "appearance" became part of the story.

But Semenya's physical presence is distinct from her gender expression. If you've ever been in a room with her you know she's not a tomboy. In addition to her height, as Levy wrote in the *New Yorker*, Semenya is "breathtakingly butch. Her torso is like the chest plate on a suit of armor. She has a strong jawline, and a build that slides straight from her ribs to her hips." Her voice is also "surprising," Levy added: "Semenya's father, Jacob, has put it" this way: "'If you speak to her on the telephone, you might mistake her for a man.'" This last detail is often remarked upon. South African sportswriter Wesley Botton remembered that after her "breakout performance in July 2009 in Mauritius," where she "came out of the blue and ran 1.56," he "tried to get hold of her and remember being very confused": "She's got such a deep voice, especially on the phone, I thought I was speaking to a man."

Because of how people react to Semenya, whenever she is sorted as "female" there are questions and sometimes there's trouble. The trouble went global after she won the women's 800 meters in Berlin. That "a man" appeared to have taken a world championship medal reserved for females set the world on fire, triggering first an official inquiry into her sex and, eventually, a battle not only within the Olympic movement but throughout society over how sex is defined and whether sports can legally continue to separate athletes in competition on that basis.

What Is Biological Sex?

Consider the physical attributes you process when you come upon someone you don't know. At least subconsciously, one of them—and more than likely the very first one—is biological sex, the binary differentiation that we name "male" and "female." Now consider the cues you use to identify someone as male or female. My guess is that you'll be hard-pressed to isolate one particular attribute or reason and that you've decided almost instantaneously based on the whole person rather than any single thing. If you work to explain to yourself what it is that caused you immediately to think "male" or "female," you'll probably land on the same attributes that trigger the typical reaction to Semenya: the person's head, face, and neck if you're close enough to see them, or if you're farther away, their whole body: its height and composition, how it takes up space, and how it moves. Depending on why you're sizing them up, you might also have used these overt cues as proxies for other physical attributes: their relative strength, speed, and power, and maybe even their chest and their genitals. The nonscientist—and even the scientist in his or her everyday life—doesn't so much define sex as intuit it from a set of external cues and visual inferences.

It turns out that our intuitive process for deriving sex—however vague it might seem—is consistent with the standard definitions. The *Oxford English Dictionary*'s definition of *sex*, for example, focuses on the male-female binary and its evolutionary purpose: "either of the two main categories (male and female) into which humans and many other living things are divided on the basis of their reproductive functions." L'Académie Française, the official guardian of the French Language, has just updated its definition of sex. The prior version read: "the organic form that distinguishes the male from the female[;] by extension, the ensemble of male or female characteristics[; u]sed collectively to signify men or women." Today, the definition is: the "[q]uality of a living being which determines whether it produces male or female gametes and goes along, for most species, with morphological or functional particularities and specific patterns of behavior. In particular: For human beings, the quality of being [or what makes] a man or a woman." The Institute of Medicine (IOM)—renamed the National Academy of Medicine in 2015—defines it as "a classification, generally male or female, according to the reproductive organs and functions that derive from the chromosomal complement." The World Health Organization (WHO) defines sex as "the different biological and physiological characteristics of females, males and intersex people, such as chromosomes, hormones and reproductive organs."

Although people often use the word gender interchangeably with sex, for scientists focused on the physical body, a distinction is essential. The WHO, for example, defines gender as "the characteristics of women, men, girls and boys that are socially constructed. This includes norms, behaviours and roles associated with being a woman, man, girl or boy, as well as relationships with others." Because gender is a social construct, the organization adds that it "varies from society to society and can change over time." The IOM explains that "gender refers to a combination of environmental, social, and cultural influences on women and men. Gender is rooted in biology but is primarily shaped by environment and experience."

Wherever it's found, the biological definition of sex—in contrast with the definition of gender—consistently reflects three facts we can all take for granted: First, humans are dimorphic: we exist in two distinct forms. Second, our dimorphism is universal and has been stable over time. Third, our dimorphism results from a set of physical sex characteristics that almost always come in twos and almost always sort in a binary pattern. Because of this, more than 99 percent of the time a child's sex as it's recorded at birth is true to their biological sex. Here are these three points again, this time with a bit more detail.

First, like all mammals, humans are "dimorphic"—meaning that we come in one of two forms. There are exceptions, which I'll discuss in a moment, but for now, it's enough to know that both the IOM's reference to the classification being "generally" male or female and the WHO's more specific reference to people who are "intersex" are nods to the tiny percentage of the global population whose sex characteristics don't all sort in the typical way; they are *not* claims about a third sex or a third type of human. Human dimorphism is less pronounced than in some other species—our males don't have the "gorgeous plumage" that fixated Charles Darwin and our females aren't giants in relation to their males like the blanket octopus—but it's still significant.

Second, our dimorphism is universal and has been stable over time. Apart from the extremely rare exception of individuals with ovotesticular disorder—what is sometimes called *true hermaphroditism*—humans all have the same reproductive sex characteristics that almost always sort in the same binary way, and the characteristics that develop from this original set themselves distribute bimodally—meaning mostly in one or another direction. This universality and stability are part of what makes us a single species.

Third, our dimorphism results from a distinct set of characteristics some of which *only* sort in a binary way and others which *usually* sort in a binary way, as either male or female, and result in two different coherent organisms whose complementary functions are reproduction and survival. As I'll explain below, sex characteristics include but aren't limited to chromosomes, gonads, sex hormones, and genitals. They work together through puberty to build the bodies we know as *male* and *female*. Definitions that appear to be limited to this illustrative set assume the knowledge that each comes in one of two types—for example, gonads are either testicles or ovaries—and are necessary parts of

9

one or the other of two reproductive systems and the different bodies that house them.

It's a political statement in some quarters today to say that biological sex isn't about reproduction, that is, about conceiving, bearing, and rearing children to the point where they can themselves reproduce; but sex undoubtedly is about this. This explains why we are one way or the other. Each of our sex characteristics exists as it does either directly or indirectly to facilitate reproduction. Our coherent forms as male and female exist at least in the first instance to the same ends. Their functional effects—including how we choose to use them—often transcend this purpose to the point where we may no longer connect them up in our minds, but that doesn't change their essential nature. For humans to make babies and survive as a species, it's not the politics that count; it's the biology, and that biology is undoubtedly binary.

What this means is that our physical traits almost always come in twos and almost always sort one way or the other. Moreover, the relationship between this natural sorting pattern and human reproduction emerges very quickly. You may remember some of this from biology class, but here's a summary of how human conception and sexual differentiation and development goes.

Every human embryo is created from a sperm and an egg. An embryo can't be created from two sperm or two eggs, or from just one of either. Eggs are large, stable gametes (reproductive cells) produced by the ovaries. Sperm are small, motile gametes produced by the testicles. Ovaries can't produce sperm and testicles can't produce eggs. Individuals either have ovaries that make eggs or testicles that make sperm.

Almost all of us have either a matched pair of chromosomes (XX) or an unmatched pair (XY), one from each parent. The parent with the ovaries can contribute only an X. The parent with the testicles can contribute either an X or a Y. Although all embryos are bipotential at the outset, meaning that they can develop as either male or female, the pair we have, either XX or XY, provides the blueprint for the subsequent development of the fetus as either one form of human—the kind that will themselves go on to have ovaries, make eggs, and contribute an X chromosome to their offspring—or the other—the kind that will go on to have testicles, make sperm, and contribute either an X or a Y to theirs.

I once guest-lectured in an undergraduate class on civil discourse at Duke University, where I teach law. The class is designed to challenge students across the political spectrum to engage critically but also respectfully on culturally fraught matters. My subject, sex in sport, was especially hot in that moment. As I was explaining the biological and equality rationales for a protected female category, one student interjected to say that none of what I was saying made sense because sex isn't binary. He said he was a biology major and had been taught that sex is simply a set of characteristics we all share, like chromosomes and gonads. He said nothing about the relationship of sex traits to one another—*why they're a set*, how they distribute, or about their ties to human reproduction. It was clear that he had either missed the lesson on human development or was simply reciting the script du jour on elite college campuses. Developmental biologists don't think of sex characteristics as belonging in an undifferentiated set because they sort predictably in a binary pattern and because, from the beginning, they work together to build one of the two bodies that becomes a person who, like this student, can make an argument and, by design, babies. That binary pattern isn't a figment of our imaginations or of nurture and enculturation. It preexists and enables these experiences.

Indeed, as we develop, the binary pattern becomes more, not less, distinct. Most immediately, within days of embryonic existence, our chromosomal blueprint, also called our *chromosomal complement*—XX or XY—drives sex differentiation in one or the other direction.

In the presence of the SRY gene, which is ordinarily on the Y chromosome, what is known as the *undifferentiated gonadal ridge*—the place where our gonads, either ovaries or testicles, will eventually grow—develops testicular (not ovarian) tissue. That testicular tissue will begin to produce sex hormones, including high levels of the sex hormone testosterone. That testosterone-based hormone system will go on to have a profound, lifelong influence on the individual's physical development and biology.

While XY humans are still in utero—a state of being that's only possible courtesy of the other human form that has a uterus—this includes the final formation of the testicles themselves, which will continue to produce sex hormones and eventually also sperm; the development of the prostate, which will produce the semen that carries the sperm to the penis; the development of the vas deferens, where sperm and semen mix; and the development of the penis, which delivers semen and sperm into the upper reaches of the vagina. That XY humans also use their penis to expel urine from the bladder to the outside world, and that some otherwise use it only for sexual pleasure (not also reproduction), doesn't alter the original reproductive design.

In the absence of the SRY gene, which is the normal state of things for XX humans, the development of the human reproductive system in utero is governed by genes, not hormones. In other words, while the development of an XY embryo and fetus is largely driven by the testicles producing testosterone, the development of an XX embryo and fetus is driven by the action of genes unmediated by the ovaries and their hormones. Thus, in an XX embryo, the gonadal ridge develops ovarian rather than testicular tissue, which matures into ovaries and the rest of the parts of that different reproductive system: the fallopian tubes, which connect the ovaries to the uterus and provide the pathways for mature eggs later in life to travel from the ovaries to the uterus; the uterus itself, where fertilized eggs will develop into the next generation of embryos; the cervix, which connects the uterus to the vagina; the vagina, which provides the passage for the penis and sperm to get to the cervix; and the clitoris, which ensures that baby-making can be pleasurable for this form of human too. That some XX humans use their clitoris only for sexual pleasure doesn't affect the original reproductive design.

It's because one thing leads to another in this way that, at birth, when the adults in the room see a baby with a discernible penis and testicles, they know that its chromosomal sex is probably XY, its hormone profile is testosterone-based, and that at sexual maturity—what we typically call *puberty*—its body will almost certainly produce sperm and develop in a physically masculine way. Doctors might say that the baby's penis and testicles are a *proxy* for these other things because they aren't independent of one another; they're a distinct system with known functions. For the same reasons, the law would say that they provide a strong *inference* or weighty *circumstantial evidence* of the existence of the rest.

If they need to know more before they make the call, for example because the genitals are ambiguous, doctors have a globally accepted *differential diagnosis* for sex. That diagnosis looks at the set of sex characteristics, including chromosomes, gonads, and hormone profile. Internal gonads are usually easy to see with ultrasound equipment, which is widely available around the world, even in underresourced locales.

The same goes for a baby born with labia, a clitoris, and a vagina. From this evidence, it's a great bet that the baby's chromosomal sex is XX, that its hormone profile will be estrogen-based, and that at puberty its body will develop a menstrual cycle and in otherwise physically feminine ways. It's also a great bet that the baby has ovaries that already contain all of the eggs it will ever produce.

Unlike testicles, which don't begin producing sperm until the onset of puberty and which continually replenish their supply thereafter—at a rate of millions per day—ovaries make all of their eggs during fetal development. From then on, they continue to shed eggs so that by puberty, only about a quarter of the original number is left. Unlike XY humans, who can, in theory, reproduce every day from puberty onward, XX humans can reproduce only about once every two years from *menarche*—the time of our first periods—to *menopause*—when our ovaries have stopped producing the levels of estrogen and progesterone necessary for fertility. Still, if this is what XX humans want to do, they usually have plenty of eggs to work with.

Up to now, I've mostly avoided affixing the words *male* and *female* to the description of how sex differentiation and development work to create our dimorphism—our two different human forms. I've done this to make the point that these characteristics, how they naturally sort and the two distinct bodies that develop as a result, don't follow from our naming them or ascribing further meaning to them—*they just are*. Theorizing sex and gender can be fun, but in the real world, we don't see set "A" and the form it builds as one thing and set "B" and the form it builds as another because we choose to. We see them this way because their differences are material, in both senses of the word: they're a physical reality and they're important. They're why we're here.

But we have language for a reason, and I'm going to start using the words *male* and *female* because they communicate important concepts effectively and efficiently. I need—and in this book I suggest that collectively we need—to retain the words in our lexicons for the concepts of biological sex and of male and female because, whether you celebrate, accept, or reject them:

- We undoubtedly exist in physically sexed bodies.

- Sex and our dimorphism are individually, evolutionarily, and politically significant.

- It's cumbersome to have to describe each set by its growing number of characteristics.

- It's insufficiently descriptive of the whole form they come together to create to use one or another sex characteristic—such as the blueprint XX or XY or the force that drives male physical development that is testosterone—to code for the rest.

- Using only one in lieu of or to code for the rest operates to erase what's not included, and that erasure causes harm. We have eons of experience with *man*, *mankind*, and *person* coding for both males and females and that hasn't gone entirely well for anyone.

For many readers, these points won't be controversial. But I recognize that not everyone will be persuaded by them. For now, all that I'm hoping for from skeptics is agreement that it matters to have the words necessary to convey important concepts efficiently and effectively, and that it's worth hearing out people like me who say that biological sex is among those concepts.

The characteristics that contribute to biological sex that develop in utero are called *primary sex characteristics*—primary because they emerge during the first period of sex differentiation and are necessary for reproduction. *Secondary sex characteristics* are those that develop through puberty—the second major period of sex development. In between, for about three to six months after birth, infants go through a period called *mini-puberty*, in which genes and sex hormones operate differently on male and female bodies. Mini-puberty extends the sexed physiological and neurological development that began in utero and is otherwise believed to prime the body for full puberty in adolescence.

Male adolescent puberty usually begins around age twelve (with a range between nine and fourteen). It includes the physical developments associated with fertility (making sperm) and sexual success (spreading sperm). Some of these are *dimorphic*, meaning they happen only in the male body. Other developments occur in both males and females, but *bimodally*—meaning differently in each of the two forms.

The pubertal developments that only males experience include the lengthening and thickening of the penis, the growth of the scrotum and testicles, the maturation and ejaculation of sperm, the deepening of the voice and growth of the larynx (Adam's apple), and changes in the topography of the face, especially along the brow ridge and jaw. Sex differences in head shape and facial features are used by forensic pathologists and anthropologists to identify the sex of skulls and also by plastic surgeons specializing in "facial femininization" for their male-to-female transgender patients.

The pubertal developments that both males and females experience, but differently, range from the obvious to the subtle: Males experience greater musculoskeletal development than females. Regional differences around the world range from 2-3 percent to 12 percent, but because females go through puberty earlier than males and males keep growing for longer, everywhere in the world males as a group are taller than females as a group. They're also on average 15-20 percent heavier than females, due to larger, denser bones and larger, stronger muscles—even though both bones and muscles can appear outwardly to be the same size. Males' energy metabolism favors muscle building over fat storage. And they have higher cardiopulmonary capacity, the result of larger hearts and lungs and higher hemoglobin (red blood cell) counts. On average, these differences result in a body with a different physiological engine, one that has more stamina and is bigger, faster, stronger, and more powerful than the female body.

Not every male grows to look or perform like an athlete. Many are less obviously tall, big-boned, and muscular than others—genetics and lifestyle contribute a lot to our physical form. But the underlying sex differences in anatomy and physiology are nevertheless there, and they explain a number of related phenomena, including how a male who is equally proportioned to a female—who looks outwardly to have the same build—can easily outperform her in the arena or overpower her on the street.

The primary mechanism for the physical changes that occur in males is a dramatic increase in the testicular production of testosterone at the onset of puberty, from about 0.25 milligrams daily prepuberty to 7 milligrams daily thereafter. Before then, starting at about age six months through childhood, males and females basically have the same testosterone levels. But at the onset of puberty, female levels remain the same and male levels skyrocket. By age thirteen, male and female testosterone levels no longer overlap.

The *endogenous* increase in circulating testosterone in males builds what we recognize as the adult male body. It's why people who want to *masculinize* or *androgenize* their form and physiology—including female-to-male transgender people (*trans men*)—take testosterone exogenously. As they're used here, *masculinization* and *feminization* are terms of art that refer not to socially constructed ideals but to genetically and hormonally driven physical development.

Female adolescent puberty usually begins about two years before male adolescent puberty, at around age ten (with a range from eight to thirteen). In females, the final development of the body to reproductive capacity includes the physical developments associated with preparing for fertility (the menstrual cycle) and sexual success (pregnancy and breastfeeding). Some of these developments are dimorphic: both males and females have mammary glands (breasts), but in females rising estrogen levels at the onset of puberty trigger the development of fat and the growth of the lobes necessary for milk production, which (in the absence of an atypical spike in estrogen) remain rudimentary in males. Only females experience the menstrual cycle, monthly egg maturation and release (ovulation), development of the intrauterine lining, and, if the egg isn't fertilized by sperm, the shedding of that lining and the passing of menstrual blood. Only females experience the shifting down and out of the hips to accommodate fetal development in utero. Females also experience changes in the topography of the face, but differently: instead of the development of the brow ridge and jaw, it's a lengthening between the nose bridge and chin and a filling out of lips and cheeks.

And again there are the dimorphic traits, those that both sexes experience but differently. Females are shorter and weigh less on average than males. Their lower weight is due to their different height, their smaller muscle mass, and the different density of their bones. Female energy metabolism is geared toward fat storage, mainly in the breasts, hips, and thighs, so that should they become pregnant, they can sustain a second body (within their own during pregnancy and at their breast after birth).

Not every female has an hourglass shape—genetics matter a lot—but even those who have "boyish" figures ("boyish" being defined as flatter-chested with slimmer hips) have some curves if their menstrual cycle is healthy.

The primary mechanism for these changes in the female body—which scientists call *estrogenization* or *feminization*—is a dramatic increase in the ovarian production of estrogen, from a more-or-less steady level below 5 picograms per milliliter prepuberty to a fluctuating (through the menstrual cycle) range from 20 to 40 picograms per milliliter during puberty. Thereafter until menopause, the range fluctuates between 30 to 300 picograms per milliliter.

This natural or endogenous increase in estrogen builds what we think of as the adult female body. It's why people—including male-to-female transgender people (*trans women*)—who want to feminize their form and physiology take estrogen and related compounds exogenously. It's also why transgender boys—like Drew Adams, who, in early adolescence, "really hated strongly the things that made [him] look more feminine; my hips, my thighs, my breasts"—sometimes take puberty blockers and male gender-affirming hormones to stop their development. Adams provided this personal testimony in a case challenging his Florida school board's requirement that students use the bathrooms that correspond with their sex at birth.

Throughout our lives, beginning in utero, the sex organ that is our brains is operating to influence our physical, emotional, and behavioral patterns in ways that are tied to the reproductive imperative. Our brains are notably plastic—meaning they're capable of changing in response to their environments—and this makes us really good at learning. It's also why it can be difficult to separate nature from nurture. Those who claim that nothing in human behavior—from maternal instinct to sexual orientation—is actually nature or natural are capitalizing on the layperson's sense of this difficulty. They're also capitalizing on one or both of the beliefs, which are more prevalent in certain cultures than others, that we're only social constructions or that—unlike our animal kin—our "will" is entirely free. But as evolutionary biologist Carole Hooven explains in her book *Testosterone: The Story of the Hormone That Dominates and Divides Us* (2021), "There is no reason to think that [genetic and hormonal] influences have been mysteriously switched off in the human lineage."

It's unethical to do the kinds of studies that would be necessary to separate nature from nurture in humans. It would be cruel, for example, to raise a child from birth in an environment that's entirely disconnected from other people in order to discover what's actually just nature. Because of this, as one neuroscientist explained to me, it's "very difficult" to prove conclusively that "biological factors cause sex differences in the human brain or behavior if you limit yourself to data on humans." But, he added, the "study of animals proves beyond any doubt that gonadal sex hormones and sex chromosome genes cause sex differences in the brain." And, like Hooven, he concluded, "there is no scientific basis to think that these biological variables would not also operate to make sex differences in humans, because of the similarity of the physiology of the research animals and humans."

This similar physiology begins at the cellular level. Like our animal kin, every cell in our bodies has a sex, and our brain cells are no exception. Brain cells are built on the same XX or XY blueprint as our gonads, they're chock-full of sex hormone receptors, and they're soaked—differently depending upon our sex and age—in those hormones for most of our lives. As with other sex-linked traits, some of these are shared and some distribute bimodally.

I'll have more to say about brain sex differences in chapter 5, but for now, among the sex-linked behavioral traits humans share is the ability from early childhood to distinguish others as male or female and to recognize ourselves as one or the other. Relatedly, by adolescence, most of us—whether we're male or female—exhibit or experience a tendency toward heterosexuality. By this I mean to whom we're actually attracted, not whom we're taught to prefer. Even those of us who aren't exclusively heterosexual—because we're gay, bisexual, or asexual—tend still to be interested in experiencing parenthood or at least in passing on our genes. We're all competitive, no surprise there, but more intra- than inter-sex—that is, females are more likely to be competitive with other females and vice versa.

Some brain sex traits distribute bimodally. Parts of the amygdala, for example, which is responsible for our responses—both good and bad—to fear and aggression, are generally larger in males. Parts of the hippocampus, which mediates emotion and memory formation, are generally larger in females. Male brains are between 9 and 11 percent larger than female brains, but we have the same number of brain cells; female brains are just denser than male brains. It shouldn't need to be said, but there's no evidence that human brain size is related to intelligence.

Related to these sex differences, although both males and females can be aggressive, males the world over are more likely to be physically aggressive. From the onset of puberty, males are more likely than females to ride solo, to be less communicative, to be risk takers, and to have higher libido. From that same point, females the world over are more likely than males to be interested in socializing in groups, in staying physically healthy and safe, in being communicative, and in engaging in empathetic behaviors like caregiving and cooperation. Beginning at puberty, when menstrual cycles effect jarring changes in the chemistry of the brain on a regular basis, females experience higher degrees of mood-related disorders including depression and anxiety, and they have double to triple the rate of migraine. Tracking their related sex-linked traits, males are more likely to suffer from substance abuse and antisocial behaviors, and to die by suicide.

We can and do civilize the noise from sex on the brain through social norms and law, self-discipline, and drugs, even to the point of being able mostly to forget it—but not entirely. While biology "doesn't lock in our reality," nature remains a powerful force. As I learned from our dog Rafa, you can breed the wolf out of the shih tzu, but as the little guy is looking to situate himself for the night, like the wolf making a bed for himself in the ground, round and round the pillow he goes. It's because nature remains a powerful force on the brain that it's a mistake to ignore biological explanations for differences in how males and females experience important endogenous and exogenous phenomena. As I write this, I'm thinking about the difference in how the sexes respond to gaming (a mostly solo activity preferred by boys with fewer ties to bad mental health outcomes) and social media (a group-based activity preferred by girls with more ties to bad mental health outcomes).

Our sex-linked biology doesn't cease to be important once we've become reproductively mature. Angelica Hirschberg, head of gynecological endocrinology at the Karolinska Institute in Sweden, explained to me, "While chromosomes, specific genes, and sex hormones together are the key players in the development of the male and female body in utero and to sexual maturity, from then to the end of our lives it's the hormones that are most significant." For females this includes ongoing menstrual cycles and their accompanying hormone fluctuations, which are sometimes interrupted by pregnancies until the cycles stop at menopause. Menopause itself is triggered by a dramatic drop in estrogen, which launches a series of changes in female physiology including but not limited to the end of fertility. Males don't have menstrual cycles and so don't experience anything close to the same degree of hormone fluctuations, or the termination of their fertility. Rather, they experience a gradual—about 1 percent per year—drop in circulating testosterone levels starting at about age forty that has corresponding implications for male physiology.

What About the Exceptions?

So far I've focused on what's generally the case, how sex characteristics sort or tend to sort in a binary way to build one of two forms of humans toward reproductive ends. In other words, I've focused on the rule. What about exceptions? For example, what about the people the World Health Organization calls "intersex"? Doesn't their existence mean we're actually trimorphic or else on a spectrum from male to female? And don't transgender people by definition transcend sex in this binary sense?

As I said earlier, what makes the rule so solid is that the exceptions don't come close to undermining it. Human dimorphism is always clear, even as we understand that certain people may be atypical in some respects. There are two reasons for this. First, the exceptions are really rare. Second, they always involve individual traits, never the entire set of sex characteristics or the body as a whole. Gender politics aside, numerically and biologically the exceptions don't result in a third type of human and they don't actually elide the differences between our two forms. We should be happy about this. If things were otherwise, we'd be an endangered species.

There are many kinds of exceptions to the binary sorting of sex traits, and they all involve sex differentiation somehow gone awry. This can happen at the genetic, chromosomal, or in utero stages of development. There can be a mutation in a gene on an otherwise normal X or Y chromosome—for example in a gene that affects the sensitivity of androgen receptors or the functionality of an enzyme that's required to process a sex hormone. The XX/XY blueprint itself can be some variation on that theme—for example an individual may have an extra X or an extra Y. Testosterone may not be secreted normally in utero, either as to time or amount.

When the deviations cause functional impairments (usually related to fertility) or health complications (like salt wasting), they're known as *disorders of sex development*, or DSD. DSD is a medical term of art that more or less corresponds to the sociological term *intersex*. Some people with DSD embrace the term *intersex*, while others reject it. Like the debate about the replacement of the terms *Latino* and *Latina* with *Latinx*, the term *intersex* isn't generally preferred by those with related conditions, because they don't have a problem with their sex or they reject the political construction of themselves as somehow in between male and female.

The incidence of DSD is truly tiny. Only about 0.02 percent of humans are exceptional as to the way their sex characteristics sort, which means that well over 99 percent of us are sex-typical. Translated into our everyday experience, this means that, overall, you have about a 1-in-5,000 chance of meeting a person with some kind of DSD. As DSD expert Richard J. Auchus explained to me, the incidence of specific conditions, including the ones you may have heard of by name—such as classic congenital adrenal hyperplasia and complete androgen insensitivity syndrome—is much smaller still.

- Classic congenital adrenal hyperplasia (CAH) is a group of diseases and the most common DSD by far. It can affect both sexes and involves the overproduction of androgens by the adrenal glands. The incidence rate is somewhere between 0.005 to 0.015 percent, meaning that overall, you have somewhere between a 1 in 7,000 to a 1 in 20,000 chance of meeting someone with CAH.

- Complete androgen insensitivity syndrome (CAIS) occurs only in males. It's characterized by the inability of the male body to read and respond to its otherwise normal testosterone levels. The incidence rate is somewhere between 0.002 and 0.005 percent, meaning that overall, you have about a 1 in 20,000 to 1 in 50,000 chance of meeting someone with this condition.

- Two other DSD that are sometimes in the news are rarer still. 5-alpha reductase deficiency (5-ARD) involves the underdevelopment in male fetuses of the prostate gland and the external genitalia, i.e., of the penis and scrotum. 5-ARD has an incidence rate of 0.001 percent. Unless you're in elite sport where athletes with 5-ARD are—in the absence of screening—overrepresented in the female category, this means you have about a 1 in 100,000 chance of meeting someone with this condition. The incidence of ovotesticular disorder or true hermaphroditism—characterized by the development in utero of some ovarian and some testicular tissue—is so small that specialists in DSD count only the actual number of recorded cases, about 500 in total to date.

When the chances of being right about a baby's sex are well over 99 percent based on external genitals, and essentially 100 percent after a complete evaluation, you can see why clinicians are comfortable making the call—and why they often disagree that, at birth, they're "assigning" rather than simply "recording" a baby's (actual) sex.

The 99 percent figure is contested by some. It's been suggested, for example, that only 98.3 percent of humans are actually dimorphic and that 1.7 percent (not 0.02 percent) are exceptional. Even if this were true, it wouldn't change the fact that almost everyone's sex traits distribute in binary pattern. More importantly, it's not true if what's being measured are conditions that affect form, function, and health, including reproductive health. The way to get from 0.02 percent to 1.7 percent is to take the set of things we characterize as DSD and add variations that have none of the same impacts. To significant disorders, it requires adding simple differences. As the organization Intersex Human Rights Australia acknowledges, increasing the number of exceptions from 0.02 percent to 1.7 percent is achieved by counting "any 'individual who deviates from the Platonic ideal of physical dimorphism at the chromosomal, genital, gonadal, or hormonal levels.'" This philosophical gambit fundamentally changes the meaning of biological sex.

The numbers would also increase if we added transgender and nonbinary people to the set, especially as they're much more numerous than people with DSD. We know, for example, that some very young children—mostly male—experience a distressful disconnect between what they see between their legs and their inner sense of their own sex. The medical term for this phenomenon is *early onset gender dysphoria*. If you want to get to know a kid with this condition, I encourage you to watch the Emmy Award–winning short documentary *Trans in America: Texas Strong*. Its wonderful subject, Kai Shappley, makes sense of the hypothesis that the condition is related to atypical brain development either in utero or during mini-puberty, so that some of what we think of as gender identity is actually brain sex. But even if this turns out to be right, describing early onset gender dysphoria as a DSD would also require buying into a fundamental change in the meaning of sex, since gender identity itself causes no alteration of the male or female form and no impairment of the reproductive function.

Exhibit A: Caitlyn Jenner. Before she came out as a transgender woman, Jenner was already famous, having won the gold medal at the 1976 Olympic Games in Montreal in the men's decathlon. The winner of the decathlon is often described as "the world's greatest athlete" because its ten events together come close to capturing the complete set of athletic skills and traits. I'm not misgendering Caitlyn when I say that pre-transition she was on Wheaties boxes as Bruce Jenner. Indeed, with her enthusiastic cooperation, *Sports Illustrated* featured Jenner on its cover twice, the first time on August 9, 1976, as Bruce, and the second on July 4, 2016, as Caitlyn adorned with Bruce's Olympic medal. In between 1976 and 2016, Jenner famously fathered six children.

Exhibit B: Freddy McConnell. A few years after he came out as a transgender man, McConnell stopped taking testosterone so that he could get pregnant. He's done this twice now and has two beloved children. Check out *Seahorse: The Dad Who Gave Birth.* It's an extraordinarily moving documentary about his experience. McConnell had a double mastectomy before he got pregnant the first time and so he couldn't breastfeed his infants. But because they are female, transgender men who haven't had this surgery can and sometimes do.

Ultimately, neither disorders nor differences of sex development affect the rule that humans are dimorphic because they don't affect sex or its purpose as a whole. With the extremely rare exception of people with ovotesticular disorder, that is, true hermaphrodites, unless a person has altered their body using cross-sex hormones or surgery, we're all male or female chromosomally, gonadally, and hormonally. And, with the very rare exceptions of males with complete androgen insensitivity syndrome (CAIS) who appear to be female and females with *untreated* congenital adrenal hyperplasia (CAH) who may appear to be male, we all appear to be the sex we actually are.

Within milliseconds we perceive—almost always correctly—a person's sex from their head and face if we can see them or from their body if that's available to us. As to the face in particular, the traits that provide us with the most information are the eyes, eyebrows, and brow ridge together, followed by the mouth and jaw together, followed by the nose. The caveat is that outside of experimental conditions—that is, in real life—we judge sex based on the whole topography of the face, on the ensemble of its features in relation to one another.

Sex seems also to determine who sees what and what we see. Females appear to be better at distinguishing sex than males, meaning that when we're on task we're more accurate more of the time; and both sexes appear to be better at identifying males than they are at identifying females. Either males are more distinct or it's more important that they be identified quickly and accurately, or both. Stanford neurobiologist Nirao Shah adds, "All social and sexual encounters are predicated on first correctly identifying the sex of the other agent," and so "[i]t's a fundamental decision animals make." Correctly identifying sex is automatic and fundamental because it's adaptive: it's good for us.

I don't know Caster Semenya's wife, Violet Raseboya, and so I won't pretend to speak to her reasons, but viewed from biology, it was understandable and adaptive for her *both* to resist when she saw Semenya come into a women's restroom *and* later to fall in love with and marry her. It was understandable and adaptive *both* for the world to be triggered when it saw a "man" in a situation set aside for "women" *and* later to come not only to understand that Semenya's circumstances are not that simple, but also to respect the way she's carried herself over the years since she first burst onto the scene as an eighteen-year-old. Given the physical differences between males and females and the significance of sport to society, it has *also* been reasonable for policymakers to set aside competition for females only *and* to do so on physical-evidence-based grounds.

For years, Semenya's allies—but notably not the runner herself—falsely accused those who believed they were seeing a male-bodied person of "ignorance," "racism," and even "misogynoir." *Misogynoir* is a bit of academic jargon: misogyny—contempt for women—specifically as it's directed toward black women. An aspect of this special prejudice has long been a social masculinizing of black women vis-à-vis white women; especially the characterization of the former as more muscular, more aggressive, and more sexual than the latter.

One writer argued, for example, that "[t]he reason for the high proportion of Black women athletes being called masculine is theorized to correlate with the misogynoir Black women face." (There is no "high proportion of black women athletes being called masculine," but the quote is otherwise illustrative.) Using gender as a synonym for sex, this author went on to accuse me of misogynoir based on arguments I had made in my scholarship in favor of sex-based eligibility standards for elite female competition:

> When Coleman explains that we want to see "female" bodies on the podium, is she saying that we want to see white women's bodies? Is she just another facet of society continuing to refuse to grant Black women femininity? Or does she genuinely believe the transphobic claim that testosterone, not one's own feelings towards their body, decides gender?

(My responses? No, no, and what a mess we've made of sex and gender!)

Another writer warned that I was couching my "misogynoir in liberal feminist language of 'equality' and 'fairness'" and then argued that my insistence on "sex equality" is "baffling" because, "[J]ust even considering how hormones fluctuate and vary among women; so, how is 'equality' possible?" (It's been three years since I first read this, and I'm still trying to wrap my head around the notion that this new feminism has adopted as its own the decidedly patriarchal point that sex equality isn't possible because female hormone levels fluctuate.)

What is clear is that neither writer did their homework on me or the many other black women who typically grace the podium in our sport: unlike Semenya's, their sex is not in doubt.

These same allies for years sought to hide the facts that would give the lie to their deception. They began in 2010 by asking sports regulators to describe *males with DSD* whose legal gender is female as *females with hyperandrogenism* or, colloquially, as *females with high testosterone*. This was a misnomer not only because the female competition category is about sex not gender, but also because the term comes from medicine, where it describes *females* whose polycystic ovaries or hyperplastic adrenals are producing more than the typical amount of the masculinizing hormone—but still much less than *males*. To be clear, the athletes who would be labeled in this misleading way are male with testicles and normal male testosterone levels. Still, the regulators, who had been beaten up in the press and on Twitter about their "sex testing" policies, wanted badly to be seen as responsive and so they obliged.

The effect of their concession was to elide the differences between the male body and the female body in cases where the person's legal gender is female. With a stroke of the regulatory pen, the category *female* came to include both males and females. Magically, testosterone levels that part ways in early adolescence could now be described as "overlapping" and any attempt to distinguish among "females" on the basis of biological sex, indeed any talk of biological sex at all, was attacked as a human rights violation.

This effort had a long and powerful tail. Going into the Rio Olympics in 2016, the *New York Times* published an editorial describing Semenya as a female with hyperandrogenism and arguing not only that she should be permitted to compete in the female category because of this, but also that, like other females with "naturally occurring abnormalities," she "deserves her shot at the gold." As you might imagine, the call by the paper of record to celebrate perfectly normal male sex traits as exceptional in women's competition was a real mind-fuck for the females in the field. Indeed, just days later, they had to step aside not just for Semenya, who won as expected, but also for two other "females high with T." The trio swept the 800-meters podium, effecting a complete erasure of the female body from the women's medals at that distance. Misogyny used to mean erasing females; now it meant refusing to celebrate that erasure.

Two years later, in 2018, the federation that governs the sport of track and field on the global stage—now called World Athletics—revised its eligibility rules for the female category. The revision conditioned the inclusion of athletes with certain XY-DSD on a period of substantial testosterone reduction: from outside to inside the female range. By its terms, the only athletes who are affected by the rule are genetic males with testicles and bioavailable testosterone in the male range.

Semenya responded to the rule change by suing the federation at the Swiss-based Court of Arbitration for Sport (CAS). Paraphrased, the question she presented to the CAS was whether a governing body can lawfully restrict a person who was assigned female at birth from competition in the "female" category on the basis of their actual sex. In the course of the proceedings, she affirmed the value of the female category—she has never asked for that to be disbanded—but she maintained that *sex*-based eligibility rules cause a variety of individual harms that amount to human rights violations for athletes whose legal gender has been female from birth. She argued that this subset of athletes has a reliance interest in that legal designation that outweighs any interests a federation might have in protecting the female category from athletes with male sex characteristics. In effect, although her formal argument excluded them, Semenya joined transgender women in arguing that there are different routes one might take to becoming, and different ways of being, a woman.

In 2019, ten years and six major medals from Semenya's 2009

World Championships debut in Berlin, the CAS upheld the federation's sex-based eligibility standards on the ground that adopting her reliance argument would be "category defeating." The podium sweep in Rio made clear that a truly tiny incidence rate can have huge impacts in sports. But the evidence of male advantage in female competition was ultimately much more extensive. Although Semenya had come to stand for DSD athletes in the public eye, the sport has had decades of experience with these conditions at podium level, including from outside of Africa and the African diaspora. For those who cared to know the facts, it was never about race.

In its reasoned decision, the CAS confirmed that Semenya is, in fact, biologically male, and that her particular XY-DSD doesn't disrupt male sexual development in the respects that matter for sport. The CAS was too diplomatic to put it so directly—by then, not only had the idea that there are such things as a male and a female body become anathema in academic and progressive circles, but also the term *biological sex* itself had become something of a slur. In one of the most brilliant judicial passages I've ever read—brilliant because of how it threads the needle between telling the truth about sex and dealing with the extraordinary politics of the situation—the CAS wrote of the need to protect

> individuals whose bodies have developed in a certain way following puberty from having to compete against individuals who, by virtue of their bodies having developed a different way following puberty, possess certain physical traits that create such a significant performance advantage that fair competition between the two groups is not possible.

That the lawyers for the federation had to say "biologically male" in their briefs to get to the point where this passage could be written is a testament to the power of the long lie *females with high testosterone.*

31

To address it effectively, one could no longer choose to be circumspect. Indeed, it had become destructive of the female category to do so: The balls are there, albeit undescended. Their location inside rather than outside the body doesn't alter their functionality. Among other things, they produce testosterone in the normal male range, a level that is ten times higher than 99 percent of females, including elite female athletes. Arguments to the contrary notwithstanding, that testosterone is bioavailable—meaning that the androgen receptors work just fine—and, since puberty, it's built a body we *accurately* read as male.

Semenya has been harmed by the scrutiny of and judgments about her body since she was a child. She's also been badly treated by those who should have been her allies and advocates. First there were the politicians who knew they shouldn't put her on the plane to Berlin but did it anyway because they wanted the medal. Then there were the strategists for the identity movement who didn't care about sports, just about having a poster child for their cause. As a result, because Semenya was offered up to these combined ends and because she kept running and winning, the rest of us who did care about sports had no choice but to deal with her on those terms. We didn't have the luxury not to talk about her body because they put her body in issue.

Since the CAS decision, Semenya has taken back her own story—including from her erstwhile allies. To her longtime chronicler at the *New York Times*, Jeré Longman, she has acknowledged again, as she did in effect to Ariel Levy in 2009, that none of this was news to her and her family. More importantly, she has come to insist that she will no longer be pigeonholed by anyone—not the regulators who insist that she take female gender-affirming hormones in order to compete in her preferred women's events and not the intersex advocates who sought to make her into a public champion of their cause. Indeed, she's recently clarified that she doesn't see herself this way at all. Semenya has come full circle. Unapologetically herself as she was before the handlers and allies got to her, she says, simply, "God made me the way I am."

Semenya's case at the CAS was the first major challenge to the traditional view that sex—as a legal and political term—means biological sex, as opposed to legal gender or gender identity. In further pursuit of this challenge, she appealed the CAS ruling to the Swiss Federal Tribunal, the country's highest court. She lost there too, on the grounds that the CAS judgment didn't contravene the law of Switzerland, including as that law incorporates international human rights norms. In the same period, South Africa successfully petitioned the UN Human Rights Commission for a ruling that Semenya's human rights are being violated, and Semenya herself appealed Switzerland's decision to the European Court of Human Rights. As I write, her case is still pending in that forum.

2

The Answer from Law

IN 1869, MYRA BRADWELL APPLIED for a license to practice law in the state of Illinois. She was thirty-eight years old, married, with two teenage children. She had trained in her husband's law office and passed the bar exam with high honors. In those days, you didn't have to go to law school to practice, you just had to pass the bar, and so no more was required of anyone.

Nevertheless, the Illinois Supreme Court rejected her application. It did so without regard to her competence (the justices acknowledged her abilities in its ruling) or her good standing in the community (equally well established). Rather, she was rejected on the ground that she was a *married female*, and then—after she petitioned for reconsideration—simply because she was *female*. In the justices' view, Bradwell's sex alone created a legal "disability" the court was "powerless" to cure. Although the licensing law was silent on the matter of sex, given that only men had ever been admitted to the bar, if a female was to be eligible, the court said, the state legislature would have to say so explicitly.

The state court justices went on to make it clear that they thought licensing females was a bad idea. Drawing from the well-worn playbook in the centuries-long debate over the Woman Question—about the physical *nature of females* and their corresponding societal role—the justices volunteered two "facts" about the *nature of males* in the legal arena: First, they claimed that the "hot strifes of the bar," the "momentous verdicts," and the "prizes of the struggle" altogether "tend to destroy the deference and delicacy with which it is a matter of pride of our ruder sex to treat" women. Second, they claimed that the "administration of justice" would suffer if women were present "as barristers in our courts."

In other words, the law is like war. In the heat of battle, male lawyers can't help but be rough and rude, and should women be injected into the arena, men—who are otherwise perfectly competent—would be distracted by "rustling garments" and "swayed"—not by the law or the evidence—but by sex, "the most powerful influence known to humanity." The justices were hardly outliers at the time. Among others, the deans of both the Harvard and Yale law schools had expressed concern about the possibility of skirts in their law libraries.

Myra Bradwell was having none of it. "Brains and mentality measured by the formation of wearing apparel. This will not do!" she retorted in an editorial. She had been raised by parents who were involved in the abolitionist movement and chose as her own cause "the freeing of women from many of the conditions of their own enslavement." To these ends, she moved forward on three fronts.

She had earlier founded the *Chicago Legal News*, which provided such valuable information and services—including the prepublication of cases and statutes—to the men of the bar that they had no choice but to take it up. In the editorials, readers were confronted with the feminist case from Bradwell and her sisters in arms. There was this, for instance, from Lavinia Goodell of Wisconsin:

> If nature has built up barriers to keep woman out of the legal profession, be assured that she will stay out; but if nature has built no such barriers, in vain shall man build them, for they will certainly be overthrown.

Then, in 1872, Bradwell drafted a bill for consideration by the state legislature. It read, simply, "No person shall be precluded or disbarred from any occupation, profession, or employment (except military) on account of sex." That same year, she also appealed her loss in the Illinois Supreme Court to the Supreme Court of the United States.

In her petition in *Bradwell v. State of Illinois*, she and her lawyer Matthew Carpenter argued that the Constitution's Fourteenth Amendment, which had been adopted just four years earlier to address the many legal disabilities that encumbered the nation's formerly enslaved population,

> opens to every citizen of the U.S., male or female, black or white, married or single, the honorable professions as well as the servile employments of life; and no citizen shall be excluded from any of them. Intelligence, integrity, and honor are the only qualifications that can be prescribed as conditions.

Or, as she put it in the *Chicago Legal News*, "One half of the citizens of the United States are asking—Is the liberty of pursuit of a profession ours, or are we slaves?"

Cognizant of the power of the claims about the nature of males that had influenced the Illinois Supreme Court, Bradwell and Carpenter sought to persuade the men of our highest court that a ruling in her favor would be good—as in civilizing—for them too:

> It will everywhere be found that, just in proportion to the equality of women with men in the enjoyment of social and civil rights and privileges, both sexes are proportionately advanced in refinement and all that ennobles nature.

As the case went on, the *Nation* declared Bradwell's arguments "ridiculous" but—this is the best part—also scientifically "interesting as showing the effect produced by legal study on the female mind."

The Supreme Court itself chose to avoid sex entirely. In an 8-1 decision against Bradwell, it found that the right to practice a learned profession isn't a privilege guaranteed by the federal Constitution to anyone. Because of this, the justices wrote, "[t]he power of a State to prescribe the qualifications for admission to the bar of its own courts is unaffected by the fourteenth amendment, and this court cannot inquire into the reasonableness or propriety of the rules it may prescribe." This is the same federalism ground that, in 2022, supported the Court's decision in *Dobbs v. Jackson Women's Health Organization* to send abortion back to the states.

One lone justice chose not to skirt the question Bradwell presented, or those aspects of the nature of females that supported the Court's ruling against her. In a concurrence, Joseph Bradley said that practicing one's profession is a privilege of national citizenship—he disagreed with his fellow justices on this point—but like his brethren on the Illinois Supreme Court, he also held strong to the traditional view that the professions were properly open only to males. In writing separately to say this, he produced one of the most famous passages in the modern history of sex in the law. I quote it here in its entirety so that you don't miss any of its patriarchal gems:

[C]ivil law, as well as nature herself, has always recognized a wide difference in the respective spheres and destinies of man and woman. Man is, or should be, woman's protector and defender. The natural and proper timidity and delicacy which belongs to the female sex evidently unfits it for many of the occupations of civil life. The constitution of the family organization, which is founded in the divine ordinance as well as in the nature of things, indicates the domestic sphere as that which properly belongs to the domain and functions of womanhood. The harmony, not to say identity, of interest and views which belong, or should belong, to the family institution is repugnant to the idea of a woman adopting a distinct and independent career from that of her husband. So firmly fixed was this sentiment in the founders of the common law that it became a maxim of that system of jurisprudence that a woman had no legal existence separate from her husband, who was regarded as her head and representative in the social state, and, notwithstanding some recent modifications of this civil status, many of the special rules of law flowing from and dependent upon this cardinal principle still exist in full force in most states. One of these is that a married woman is incapable, without her husband's consent, of making contracts which shall be binding on her or him. This very incapacity was one circumstance which the Supreme Court of Illinois deemed important in rendering a married woman incompetent fully to perform the duties and trusts that belong to the office of an attorney and counselor.

It is true that many women are unmarried and not affected by any of the duties, complications, and incapacities arising out of the married state, but these are exceptions to the general rule. The paramount destiny and mission of woman are to fulfill the noble and benign offices of wife and mother. This is the law of the Creator. And the rules of civil society must be adapted to the general constitution of things, and cannot be based upon exceptional cases.

The humane movements of modern society, which have for their object the multiplication of avenues for woman's advancement, and of occupations adapted to her condition and sex, have my heartiest concurrence. But I am not prepared to say that it is one of her fundamental rights and privileges to be admitted into every office and position, including those which require highly special qualifications and demanding special responsibilities. In the nature of things, it is not every citizen of every age, sex, and condition that is qualified for every calling and position. It is the prerogative of the legislator to prescribe regulations founded on nature, reason, and experience for the due admission of qualified persons to professions and callings demanding special skill and confidence. This fairly belongs to the police power of the state, and, in my opinion, in view of the peculiar characteristics, destiny, and mission of woman, it is within the province of the legislature to ordain what offices, positions, and callings shall be filled and discharged by men, and shall receive the benefit of those energies and responsibilities, and that decision and firmness which are presumed to predominate in the sterner sex.

In other words: Men and women occupy separate "spheres" and have separate "destinies" because nature and God dictate that it be so. Our gendered norms and laws, our social constructions, derive from these original sources. The occupations of public life are the domain of men. The role men play also includes "protecting and defending" women—who are naturally "timid and delicate"—from the inherently masculine "energies" required by these tasks. Men are not only stronger in body but also stronger in mind. Compared to women they are decisive, "stern," and "firm." The sphere of women is private or "domestic" life. The roles women are to play are those of "wife and mother" with their associated household functions. The concept of the family as a unit with the man and woman playing complementary roles—man as its public face and woman as its private face—doesn't allow for the possibility that a woman might also step into a public-facing role separate from that of her husband. This concept is so well cemented that traditionally, once married, a woman had no separate legal identity. Unmarried women are exceptional, and social norms and the rule of law can't be based on exceptions. They must be based on what's usually true, that is, on "the general constitution of things."

How did we get from Myra Bradwell is female to Myra Bradwell is a woman, wife, and mother, and so wears skirts and can't be a lawyer, while Joseph Bradley is male and a man and so can be a lawyer, but won't be a good one if he's distracted by Bradwell's skirts? What has the law meant—and what does it still mean—when it uses the words *sex, male* and *man, female* and *woman?*

What Is "Sex" in Law?

In chapter 1, we saw that the field of biology defines the word *sex* based on its goals and what it's trying to communicate. That is, because biologists seek to understand the nature and functions of living organisms through the life cycle, they generally define *sex*—and its associated terms *male* and *female*—to mean, per the Institute of Medicine, "a classification, generally male or female, according to the reproductive organs and functions that derive from the chromosomal complement." Those aspects of our biology, including the structures and systems that we know flow from our chromosomes and gonads, are a piece with how we understand the word *sex* given the goals of the scientific enterprise. Biologists also distinguish *sex* (the biology) from *gender* (the social construction) because, while both can affect health and well-being, they aren't the same thing and they don't do the same work.

The law is like biology. It too defines the word *sex*—and its associated terms *male* and *female*—based on its goals and what it's trying to communicate. The reasons sex is significant to the institution become part of how we understand the words in legal settings.

The law's overarching goal is to organize and regulate people and societal institutions so that the community's basic needs are met. Laws that are designed to protect people from physical violence, to secure food and housing, and to facilitate fertility are good examples. The first two are obvious, but the third may not be. That is, the idea that the government should be involved with fertility may be anathema to those who think of sex, reproduction, and childrearing as purely private matters, but the notion is far from uncommon. All societies keep tabs—rough or refined as these may be—on the numbers of live births, of children surviving to the age when they can reproduce, and the sex of those children. They do this because without people there is no society and without reproduction there are no people. Beyond this bottom line, it is citizens in their working years who support (directly or indirectly through taxes and pension plans) the youngest and oldest members of the community.

Once societies get beyond the subsistence level and have the luxury of choice, the law's role begins to include regulating according to prevailing cultural norms and aspirations. Laws that are designed to secure liberty and equality are good examples here. So too are those that prescribe acceptable body covering—like the nudity law in California that prohibits "exposure of the genitals or buttocks of any person, or the breasts of any female person" and the law in Iran that makes the hijab compulsory for "women who appear in public."

The combination of the law's two roles—regulating for the basics and regulating according to prevailing norms and aspirations—is what

41

Joseph Bradley meant when he wrote in his concurring opinion in *Bradwell v. Illinois,*

> It is the prerogative of the legislator to prescribe regulations founded on nature, reason, and experience for the due admission of qualified persons to professions and callings demanding special skill and confidence. This fairly belongs to the police power of the State.

In U.S. law specifically, the *police power* is the authority of the government to regulate health, safety, and morals in the interests of the community's general welfare. Whatever this source of governmental authority is called, every country has a legal system—whether it's customary or codified, secular or religious—that's designed to meet these three goals.

Because sex and sex differences matter for the health, safety, and morals of the community, sex itself has always been one of the first ways in which people are sorted as humans organize themselves and establish rules for their societies. There are many other traditional bases for classifying people, including age and infirmity and in- and out-group markers like caste, race, and tribe or ethnicity. But consistent with the fact that our social encounters are typically "predicated on first correctly identifying the sex of the other agent," sex has always been a primary focus within this set.

Health is connected to sex because of the biology of human reproduction and aging. The sustenance of infants to the point where they can survive without breast milk requires very different contributions from females and males. It's true that we now have a suite of tools—laws, pharmaceuticals, surgeries, surrogates—that allow us to imagine that we're beyond nature, but these tools are unprecedented in human history and only available to a sliver of the population even now in the United States. Exceptional circumstances don't speak to the lives of most people. Health is further connected to sex because human development and many of the conditions and diseases associated with aging track sex and are affected by biological sex differences. I'll have more to say about this in chapter 5.

Safety against physical violence has always been connected to sex because males are *both* the principal perpetrators of individual crimes, of gang violence, and of military aggressions, *and* the community's first protectors and rescuers, as police officers, firefighters, and soldiers. That females can also be violent and an integral part of police forces, fire departments, and the military hasn't altered this overwhelmingly male story. Again, exceptional circumstances don't speak to the lives of most people and—Justice Bradley was right about this—rulemakers are charged with the general welfare.

Finally, morals have always been connected to sex because culture and religion—the primary sources of a society's values-driven laws—tend to develop around basic needs. This means that lots of our norms and aspirations are tied to (having) sex, fertility, raising kids, and sustaining and protecting the family and the community. Morality in law is not just about sex-based clothing and privacy. Taking today's culture wars as an example, abortion, marriage, school curricula, book bans, gender-affirming care, and parental rights are all tied to sex.

From these contexts, it's clear that the law has always defined *sex* according to our *nature* and *natural characteristics*, and that it has used the word as we do in everyday speech, to mean the male and female body. Two formal definitions reflect this consistent usage over time, both from *Black's Law Dictionary*, the standard legal reference in the United States. In the first and second editions, sex was defined as:

> The distinction between male and female; or the property or character by which an animal is male or female.

Since its third edition in 1933, sex has been defined as:

The sum of the peculiarities of structure and function that distinguish a male from a female organism.

The eleventh edition, published in 2019, contains this same definition. Several U.S. states have recently changed their definitions of sex in response to lobbying by trans rights organizations, retiring the established focus on the male/female binary and the whole body or "organism" in favor of gender identity and a nod to the characteristics that "in sum" make up sex. But in American law generally, *sex* still means *biological sex*.

The most important example of this is the Supreme Court's 2019 decision in *Bostock v. Clayton County. Bostock* affirmed that it was a violation of Title VII of the Civil Rights Act of 1964—which prohibits discrimination in employment on the basis of sex—for a funeral home to fire a transgender woman simply because she was transgender since doing this required taking into account her (male) sex, which was irrelevant to the job. In reaching this result, which guarantees that transgender people have the same rights in the workplace as everyone else, the Court declined to redefine sex away from biology to be or to include gender identity. Following *Bostock*, a federal appellate court applied Title VII's exception for circumstances where sex *is* relevant to the job ruling that a transgender man who works as a prison guard is not male for purposes of the prison's same-sex strip search teams. In doing so, it found in favor of a Muslim prisoner whose religion doesn't allow him to be seen or touched unclothed by someone of the female sex, whatever their gender identity.

Another important example: The United States requires all male residents and citizens to register with the Selective Service System when they turn eighteen. The law is designed to ensure an accurate recording of all eligible males in the event of a future military draft. As I write, this law doesn't make an exception for transgender women, although like others whose biological sex is male who may have good reasons not to serve, it allows them to petition to be relieved of their obligation on a case-by-case basis. Earlier conscription laws, including those of the Union and Confederacy during the Civil War, were also restricted to males. Females did serve in the Civil War, but they did so in noncombat roles or, exceptionally, by passing as male.

One final example: The National Institutes of Health (NIH) require that applicants for government grants for medical research take sex into account in their work. The goals of this requirement are to ensure that female subjects are included in the research, and that work is done to understand the nature and extent of the biological differences between males and females so that federally funded medical research also benefits the female half of the population. As we saw in chapter 1, coming from science this requirement explicitly distinguishes *sex* as biology from *gender* as social construction.

In contrast with the definitions and usages from science, since 2009, *Black's Law Dictionary* has added that in law, *sex* can also mean *gender*. You know this mostly as a complete synonym for *sex*, for example when you're asked on forms to mark your *gender* as male or female. This change is often credited to Ruth Bader Ginsburg, who was one of the principal architects of modern sex discrimination law. Ginsburg herself credited her "astute secretary" Milicent Tryon, who told the future justice,

> I'm typing all these briefs and articles for you and the word sex, sex, sex is on every page. Don't you know that those nine men when they hear that word, their first association is not what you want them to be thinking about? Why don't you use the word gender?

What that one-word substitution, *gender* for *sex*, didn't change was the ubiquity of the words *woman* and *man* as synonyms for *female* and *male*. As you know from your own life experience, the words *woman* and *man* were, and mostly still are, used to mean "an adult human female" and "an adult human male." As Justice Ginsburg put it in her majority opinion in *United States v. Virginia* (1996), "The two sexes are not fungible . . . Physical differences between men and women are . . . enduring" and "immutable."

The law has also traditionally sexed related words like *spouse* and *parent*. Although this practice has been wound down significantly in recent decades to take account of same-sex couples, it remains a factor when reproduction is at issue. Case law and statutes on the rights of unmarried *fathers*, for example, focus on males whose sexual relations with a female result in the conception and birth of a child. Males in this category are not just "sperm donors." Correspondingly, vital records laws that describe the required contents of birth certificates focus in the first instance on *mothers* as the *female* or *woman* giving birth.

The California Health and Safety Code is illustrative here. It requires that a "certificate of live birth" be filed with the government that includes the sex of the child, the name of the *mother*—who is defined as *the woman giving birth*—and that individual's last normal menses and pregnancy history. If it's a surrogacy or an adoption situation, the parent(s) who will raise the child may be uncomfortable with the legal term *mother* as applied to the female who gives birth. They may prefer language like "gestational surrogate," which disconnects the pregnancy from parenthood as "sperm donor" does. Or they may prefer that this history be erased from the record book entirely.

But as the Family Division of the High Court of the United Kingdom noted in a recent case brought by a transgender man who gave birth and wanted to be listed as *father* not *mother* on his child's birth certificate, this misunderstands the document and its public purpose. Even as the birth certificate contains information about the parent, it belongs not to them but to the child and to the state, both of whom have separate (from the parent and each other) interests in an accurate recording of the facts about the child's gestation and birth. As that court explained it, for purposes of the child's birth record:

- "There is a material difference between a person's gender and their status as a parent."

- "Being a 'mother' whilst hitherto always associated with being female, is the status afforded to a person who undergoes the physical and biological process of carrying a pregnancy and giving birth."

- "Whilst that person's gender is 'male,' their parental status, which derives from their biological role in giving birth, is that of 'mother.'"

I've already introduced you to the petitioner in this case—he's Freddy McConnell, whose documentary *Seahorse: The Dad Who Gave Birth* I recommended in chapter 1.

On Reasonable and Unreasonable Inferences

Although the law has always used the word *sex* to mean *biological sex*, as Myra Bradwell's story makes clear, the legal definition of the word packs in more than just the biology. The words *sex*, *male* (and *man* or *boy*), and *female* (and *woman* or *girl*) have also been defined according to inferences we make from what we know about the biology.

As we saw in chapter 1, making inferences is automatic and adaptive. It's part of how we make sense of our world from the information we have. You walk into a room and see a lone man holding a smoking gun and you infer from this that he just shot the person lying dead at his feet. I look at the sky and see gathering clouds, and I infer that it might rain and pack an umbrella. Many inferences are fine, either because they're rational given the available evidence or because they turn out to be accurate. Others aren't, either because there's ample evidence in the moment that they're just plain dumb or because new evidence shows they were inaccurate. It was plausible to think that the sun revolves around the earth until we learned it wasn't. Assuming that a young black man dressed in athletic gear jogging down the road is in the neighborhood to steal rather than simply to exercise is dumb and dangerous.

The inferences we've made over time about sex, the nature of a woman (what it means to be female), and the nature of a man (what it means to be male) are no different: they can be fine, dumb, or no longer plausible.

It was and remains fine, for example, to infer that people who appear to be female at birth will likely feminize physically at puberty and experience a menstrual cycle with some predictable brain effects, namely, those that are typically associated with premenstrual syndrome (PMS), cyclical migraine, anxiety, and depressed mood. It was and remains fine to infer that people whose bodies feminize at puberty are generally not as strong, fast, or powerful as people whose bodies masculinize at puberty. These were always reasonable inferences to make from what we knew and it turns out that, as a general matter, we were right.

By contrast, it was dumb to leap from these well-informed inferences to ideas that women are physically "delicate" and "rationally defective" in no small part because there were always many females—not just an exceptional few—who did physically strenuous work and who were as smart as the males in the room. That they existed didn't make them male or masculine or even men trapped in a female body—although this was sometimes the suggestion. Females who defy false stereotypes have always existed.

Dumb too were the pseudoscientific elaborations on these weak inferences, such as this gem from an 1842 medical school graduation speech on the hot topic "Some of the Distinctive Characteristics of the Female":

The "timid" female brain "grasps with difficulty" "subjects such as civil engineering and integral calculus, and other mathematical studies" to the point where the extra energy required of females undertaking such studies "influence[s] materially the nervous system primarily and secondarily the generative organs."

You read that right. Not only our nerves but also our *generative organs*—aka our reproductive systems—would suffer from the rigors associated with learning math. Lest you think this some outlier, just some random dude at one medical school, check out Harvard's Edward Clarke, who wrote extensively on the hypothesis that both males and females have finite energy, that pursuing higher education at elite academic institutions depletes that energy, but with damaging effects only on the female reproductive system. It turns out the sentiments expressed by the good men of the *Nation* in 1873 are stunning only out of context.

Males were also subject to the range of inferences from fine to dumb about their natures. It was fine, for example, to infer that people who appear to be male at birth would likely masculinize physically at puberty and be stronger, faster, and more powerful in general than females. It was also fine to infer that there is a natural (biological) basis for the male tendency to protection and aggression, higher libido, and sexual distractibility. These were always reasonable inferences to make from the available evidence, and it turns out that in general we were right.

By contrast, the leap from these to males can't be civilized was dumb, as were the assumptions that greater physical strength, the lack of a menstrual cycle, and a bigger brain lead to greater cognitive capacity. As the seventeenth-century philosopher Mary Astell quipped in one of her treatises, "'Tis only for some odd accidents, which philosophers have not yet thought worthwhile to inquire into, that the sturdiest porter is not the wisest man."

The strained set of inferences about the superiority of the male brain led to the conclusion that, as compared to females, males are *natural* leaders, and this was always a house of cards. Here what was most interesting scientifically was not the effects of intellectual pursuits on the female brain, but rather the effects of patriarchy on the male's.

On Cultural Artifacts

Beyond biology and the inferences we make about the nature of women and men, there are the cultural expectations we attach to each of the two groups ostensibly because of their natures. These raise questions, such as:

- Whether males can wear cosmetics and no-legged garments, including robes, dresses, and skirts; and whether females *must* wear skirts. It blows my mind that we went from "skirts are too distracting" to "women must wear skirts," but that actually happened.

- Whether males can hold hands. Definitely not in Texas, except when it's George W. Bush at his ranch in Crawford in 2002 and 2005 with Crown Prince (later King) Abdullah of Saudi Arabia. As the Saudi embassy spokesman said at the time, "When you hold somebody's hand and you lead them, it does show great respect and admiration."

- Whether males cry at all or in public. The *Harvard Business Review* in 2013 led with: "Nadal Is Strong Enough to Cry. Are You?" The *India Times* was on the same case with this headline in 2022: "'Real Men Get Emotional': Rafael Nadal Crying for Roger Federer Leaves Internet Heartbroken."

The leap from sex to inferences about sex to cultural artifacts we attach to people based on their sex takes us from *male* to physically *masculinized* to socially *masculine*. It take us from *female* to physically *feminized* to socially *feminine*. As the examples above show us, what it means to be socially *masculine* or *feminine* is different depending on the time and place because these are cultural propositions. There are lots more examples of the social construction of masculinity and femininity—of societies taking the same biology and attaching to it different behaviors and roles—but the most historically consequential is patriarchy.

If we're honest, we know it's true that nature distinguishes between males and females. We also know that without significant intervention, these natural distinctions result in a version of different "spheres and destinies" for each. In the United States before the Civil War, for example, the "average enslaved woman gave birth to her first child at nineteen years old, and thereafter, bore one child every two and a half years." At the same time, the "typical American woman" living in freedom "bore an average of 7 children. She had her first child around the age of 23 and proceeded to bear children at two-year intervals until her early 40s." In the in-between years, both sets of women breastfed their youngest. Infant "formula"—it really is an extraordinary concoction if you think about it—would eventually disconnect many females from breastfeeding, but it wasn't a thing back then and still isn't in many parts of the world. As we learned in 2022 during an unusual formula shortage in the United States, access can become an issue even for privileged people in the developed world. Because many American women with babies depend on formula to live modern lives—and our economy depends on their workforce participation—resolving the shortage ultimately involved the president himself. Women who have more control over their sex lives have altered this trajectory, but regardless, men's time was and remains otherwise occupied. That's nature.

What's undoubtedly artifact are the notions that females belong only in the house and males belong only outside of it: these are exclusively the product of certain cultures and religions and the preferences of those who have political power.

When cultural and religious artifacts from dress codes to behavioral norms to patriarchies (and matriarchies) are embedded in law, they affect the way we read—and what we mean by—the operative words. As Joseph Bradley's opinion in Myra Bradwell's case illustrated, in their time and in the law, the words *male* and *man* coded for strong, stern, steady, pants, professional, leader. The words *female* and *woman* coded for wife, mother, timid, delicate, dependent, skirts.

The law has come a long way in the century and a half since the *Bradwell* decision. Huge strides have been made to ensure that this institution is working only with real (not manufactured) and immutable (biological) differences. The best evidence of the success of this effort is probably the *Bostock* decision. Its effect was to strip from the law's idea of *male*, not the biology, because that's always been the through line, but the blind inferences and harmful artifacts that suggest it's somehow unnatural to have a nonconforming *inner* sense of oneself and a correspondingly desire to live a concordant *external* life. In its place, the Supreme Court left a more minimalist definition of *male* and of *sex* that stands in stark contrast to the ornate versions of the past.

In October 2020, a century and a half after Myra Bradwell appealed the denial of her law license to the United States Supreme Court, a pregnant Brianna Hill, also of Chicago, went into labor while she was taking the Illinois bar exam, at home, at the height of the Covid pandemic. Unlike Bradwell, Hill didn't fulfill her obligations to womanhood before she asked to be a lawyer: she went for both at the same time. She completed the first part of the exam sitting on towels because her "water"—the amniotic fluid that surrounds the fetus while it's in utero—broke before she was taken to the hospital. Less than twenty-four hours later, after she had labored to give birth to her child, she took the second part of the exam, at the hospital, sitting on an ice pack. As she told the *Above the Law* blog a few days later, "Going into labor really put the bar exam into perspective and made my nerves go away really quickly." Her story went viral twice: the first time when she took the exam and the second when she learned she passed.

Hill's story had that viral effect because in the literal blending of the bar exam with motherhood—including in all of the physical and physiological processes that so obsessed the legal and medical patriarchs back in the day: the nerves, the hormones, the fluids, the physical labor, and, given the reference to an ice pack, probably also an episiotomy—it answered the centuries-long Woman Question by completely destroying the unreasonable inferences and cultural artifacts on which their edifice was built. *Supermom* was in Hill's viral word cloud, and that made perfect sense. Other words that ran on a loop on social media—*brave*, *capable*, and *strong*—were literally the opposite of those Joseph Bradley and others of his ilk had used to define *female* and thereby legally to handicap a woman's prospects in society and law.

Because unreasonable sex-based inferences and artifacts are often complementary, time has also destroyed many of those that encumbered the words *male*, *man*, and *father*. Beyond Rafael Nadal's tears and fathers who "mother" infants, there are men everywhere who, in both real and metaphorical battles, treat women with care and respect—in war zones, in courtrooms, and yes, every day all the time in law school libraries. As Myra Bradwell suggested in her brief to the Court, it causes harm when we assume that because male behavior has a biological basis, the words *male* and *man* code for uncivilized.

While the Supreme Court rejected her legal argument—it would take another hundred years and the incomparable Ruth Bader Ginsburg to convince the justices that the Fourteenth Amendment applies also to prohibit sexism—Bradwell was successful with the Illinois legislature, which passed a bundle of laws that would do on the local level what the U.S. Supreme Court wouldn't do at the time on the national.

Her protégé Alta Hulett was admitted to the Illinois bar the same year that Joseph Bradley wrote his ode to the patriarchy. In addition to the law that prohibited discrimination on the basis of sex in occupations, professions, and employment, Bradwell can been credited with Illinois's original married women's earnings act, one of its married women's property laws, a law giving mothers equal custody rights, and an equal liability—or rights and responsibilities—law. Along the way, she continued to use the *Chicago Legal News* to cause good trouble, like writing editorials demanding that Mary Todd Lincoln be freed from the institution to which her son Robert had her committed following the president's assassination. Bradwell was successful in that effort too, and this ended up contributing to the enactment of a law that made it harder for men to commit their mothers, wives, and daughters to mental institutions.

The courts themselves did come around to Bradwell in the end. In 1890, on its own motion, the Illinois Supreme Court granted her a license to practice law, backdating it to 1872, the year she originally applied. Two years later, the U.S. Supreme Court made her a member of its bar on a motion filed by the attorney general. Bradwell never argued a case before our highest court but, still, you'll find her name on the first page of the clerk's list of "Lady Lawyers." She died two years later, not as the woman who lost *Bradwell v. Illinois* and to whom Joseph Bradley's utterly inapposite words attached, but, in those of her biographer Jane M. Friedman, as "one of the few female tycoons during the post–Civil War era" due to "the *Chicago Legal News*'s economic and journalistic success," which "served as a means, the vehicle by which [she] waged successful warfare on so many of the legal and social inequities of her day."

3

The Answer from Progressive Advocacy

H ISTORY WAS CALLING DURING Ketanji Brown Jackson's 2022 confirmation hearings to be the 116th justice of the United States Supreme Court. All nominees are history-making in some sense as they sit on the bench for life and make decisions that impact every American. We know from common sense and experience that it matters who they are and what they believe. Jackson was just the third black person ever to be nominated to the high court—the other two being Thurgood Marshall in 1967 and Clarence Thomas, who replaced Marshall in "the black seat" in 1991. Before Marshall and Thomas were appointed, a total of 113 justices had debated case after case about whether African Americans—not people of color generally but specifically African Americans—were "persons" and "citizens" under the Constitution, whether they had rights at all, whether theirs were equal to those held by others, and then what it meant to say they are—all without a single black person in the room. With Jackson's confirmation thirty years after Thomas's, there were two historical firsts: the first time the Court would open up a second seat for a black jurist and the first time it would have one for a black woman.

This second "first" was a big deal for the law and for black women and girls across America. It was an especially big deal for black women and girls who are, like Jackson, descendants of American slaves. For centuries, American law treated women and girls with African blood differently from everyone else. Enslavers had the legal right to use them for three kinds of involuntary, unpaid labor. Most familiar was ordinary servitude, albeit in inhumane conditions—in fields, workshops, or the home. Less familiar was sexual servitude, which, being at the pleasure of the master in circumstances where they lacked the legal capacity to refuse or consent, we now call rape—their African blood distinguished black women from others whose womanhood was protected by the criminal law. Least familiar to us now but commonplace at the time was the use of black women and girls for childbearing that improved the master's bottom line.

Enslaved black women's children belonged to their enslavers to use, keep, sell, or bequeath just as much as the "increase" of their pigs and cattle. The legal doctrine that allowed this is called *partus sequitur ventrem*, which means the status of the child follows that of its mother. A child birthed by an enslaved woman was born in bondage, whereas a child birthed by a free woman was born in freedom. All regardless of the father's status. In the United States, the doctrine stems from the 1662 decision by the Colony of Virginia to change, *for people with African blood*, the common law rule that the status of the child follows that of its father. In so doing, Virginia and the states that followed its lead rejected the merits of freedom claims brought by mixed-race children whose fathers were white. For example, my ancestor Ellen Craft was the daughter of her white master, James Smith, and his mixed-race slave Maria.

Partus as it applied in Georgia ensured that Ellen was born enslaved and that Smith's race was irrelevant to her legal status. *Partus* was critically important for everyone who counted directly or indirectly on slave labor, especially after 1809, when it became illegal to import Africans into the country. Enslaved black women bearing children became the only source of new free labor in the country, which, until the Civil War started, depended on southern cotton for its global economic position.

This brutal legal regime was propped up by two sets of coordinated fictions. The first was that *by their nature* females with African blood are not inclined to be intelligent and educable, or autonomous and self-respecting. The second was that *by their nature* females with African blood are inclined toward menial and physical labor, to be sexually promiscuous, and not to care as much as other mothers about the children they bear. These fictions were convenient for those who conceived them, of course. If you could pick a set of traits that would justify the servitude forced on black women and girls because of their sexed bodies—and the wholly different legal treatment of white women and girls—you couldn't be more precise than this. It's part of what makes racism in America "structural" (meaning in our bones) that these fictions have persisted in one form or another beyond the period in which they were designed to be maximally useful. The fictions have persisted even as they've been repeatedly falsified by the oral traditions in families that remember the pain caused by rape, family separation, and educational neglect. And they've persisted even as they've been repeatedly falsified by black women themselves, by who they actually are and what they actually do—when we don't obscure them from view.

When Jackson appeared before the Senate Judiciary Committee for her confirmation hearings in March 2022, she embodied the irrefutable case against this pernicious collection of lies. First, there was her name, Ketanji Onyika—deriving from a West African language and meaning "lovely one"—which reflects the gift and treasure her family saw in her birth. Then there was her tight-knit family: her parents and their fifty-four-years-long marriage alongside her husband and their own two children. There were the loving exchanges with her adolescent daughters, Talia and Leila. There was her education, her experience, and her obvious excellence: magna cum laude from Harvard-Radcliffe College, law review at Harvard Law School, Supreme Court clerk, appellate litigator, public defender, federal district court judge, and appellate court judge. There was her intelligence, her professionalism, her warmth, her dignity, and her vulnerability: all of this shone throughout the hearings.

Senate confirmation hearings are a constitutional requirement—in our system of separated powers it's not only a right but also the obligation of senators to vet rigorously the president's choices to ensure that nominees are qualified for their roles. But in Jackson's case they were also a celebration. The celebration was not of a jurist's personal and professional accomplishments without regard to her race and sex, but *because* of her race and sex. As New Jersey senator Cory Booker—himself only the seventh black person elected to the Senate—noted in his "joyous" introductory remarks on the first day, many resisted this point as an injection of identity politics into a process they felt should be race- and sex-blind. But given the particular history and the role of the Court in the sustenance of slavery and Jim Crow—the post-Civil War "separate but equal" regimes—this was neither right nor possible. Harking back to that history, which had also served for centuries to deprive the country of the true talents of black women, Senator Booker noted that we were seeing

[s]omething we've never seen before. Judge Jackson's nomination breaks an artificially confining mold of our past and opens up a more promising potential filled future for us all as Americans.

A photograph of the hearings went viral, taken on the first day by Sarahbeth Maney of the *New York Times*. In the photograph, Jackson is in the foreground, out of focus, and her daughter Leila is to the side in the background, in focus, beaming at her mother. In a story about the moment she captured the shot, Maney—who is also black—explained,

> I just remember seeing Judge Jackson smiling a lot, and I think she was receiving compliments and praise. . . . And then I noticed how proud her daughter was of her, and it gave me chills when I saw this look that her daughter gave her. It was just this look of such pride and admiration.

It was a look so many of us had on that day and throughout the hearings. A woman like Ketanji Brown Jackson was no longer just an object of law. And she was no longer merely its subject. Now she was also its maker.

The first day's sense of celebration was quickly muted, though, by Senator Marsha Blackburn, a white woman from the former slaveholding state of Tennessee. Blackburn asked Jackson if she could "provide a definition for the word *woman*." It felt like a hijacking—twice over.

Forgetting the details of slavery and its justifications—including the special dehumanization reserved for black women—is a national habit. And so it's entirely possible that neither Blackburn nor her staff considered this history when they prepared their version of the Woman Question. In fact, given the history we're not taught, it's entirely possible they didn't know that white women had joined in constructing the nineteenth-century version of womanhood that endowed them with precisely the opposite set of traits to black women, who were masculinized in the process: where black women were sexual, white women were chaste—which is why a black woman could be raped with impunity and a white woman could have a black boy lynched just for looking at her. Where black women were callous about the children they bore, white women were paragons of maternal virtue—which is why separating black women from their children was of no moment but separating white women from theirs was inhuman. Where black women lacked dignity and education and were physically strong, white women were modest and cultivated and physically weak—which is why black women belonged in the fields while white women were protected from manual labor. It was in part this fictional set of opposites that animated Sojourner Truth's famous rhetorical question, "Am I not a woman?"

I don't know whether Blackburn considered the history, but we do know that her focus was elsewhere at the hearing, on transgender women like former University of Pennsylvania swimmer Lia Thomas, whose inclusion in female competition is justified by certain progressives on the claim that "trans women *are* women" without regard to their sex. What the senator wanted to know was Jackson's view of the embedded argument that sex isn't about the male and female body after all, so that, all of the history and hormones notwithstanding, a person doesn't have to be *female* to be a *woman*. In asking the question, Blackburn was joining a Republican initiative to position the party as the protectors of *women* in the traditional sense of the word against progressives who have hijacked not just women but "especially black women" in service of their movement.

This last phrase—"especially black women"—is everywhere in progressive discourse. It's partly reparative, at once an acknowledgment that the women's rights movement historically sidelined black women and a "lifting up" of them now. It's also opportunistic, as it capitalizes on the unprecedented goodwill shown toward black women in this period to garner support for entirely different agendas.

The clearest example of this is the topic with which Blackburn was most immediately concerned: girls' and women's sports. As progressive advocates work to dismantle sex-based eligibility standards for that institution—which mainly involve checks for birth sex and testosterone levels—they tap into the politics of policing and abortion. Thus they describe these checks as "policing the female body" and they claim that this is "bad for all women" but "especially for black women," because, they say, black women are more likely to be seen as masculine.

I described a version of this claim from inside academia at the end of chapter 1; here, in an essay for NBC News titled "Caster Semenya Is Being Forced to Alter Her Body to Make Slower Runners Feel Secure in Their Womanhood," is the American Civil Liberties Union's Chase Strangio—himself a white transgender man—importing that discourse into his political advocacy. Among the "slower" runners he disses in the process are the most talented, best-trained black females on the planet:

> In other words, the CAS and the Swiss appeals court have decided that differential treatment for Black women, trans women and inter-sex women is required for athletic competitions to be "fair" to other women—at least, it is under a system in which white people wield tremendous power over the bodies and autonomy of those who are perceived to be a threat. These decisions come during the current political moment of global attack against individuals who do not fit stereotypes of binary sexual difference, and after a long history of white authorities policing the bodies of women of color, particularly those Black and Indigenous women from the global south.

And here is the National Women's Law Center, whose legal director was my antagonist at the House Judiciary Committee in 2019, arguing against West Virginia's ban on transgender girls in girls' sports, as that ban was to be applied to a white transgender girl:

> The enforcement of anti-trans sports bans often include dangerous and unscientific "sex testing" schemes that create new risks of sex harassment and assault against student athletes, which can involve anything from collecting sensitive medical documents to needless, traumatizing genital examinations. Athletics bans especially target

Black and brown women (who face increased body policing and gender scrutiny based on racialized stereotypes of femininity) and intersex women and girls (who are born with natural variations in sex-linked characteristics).

This line of argument isn't only dumb—as in ignorant of the facts—and misogynistic—because it subordinates females to males and belittles them in the process. It's also itself racist because unless these academics and advocates are writing it in blog posts, press releases, and briefs, no one—especially no one in sports—is actually thinking that black women are male and sex-testing them to prove it. The claim "especially black women" here isn't in support of black females who are the subject of their false and damaging stereotypes and who benefit as much as anyone from the existence of separate-sex sport. It's rather in support of broadening the category *women* beyond sport to include certain people who aren't female—regardless of their race. Exhibit A, the American who has been subject to the most intense "policing" of all: Lia Thomas, a very tall, broad-shouldered, flat-chested, slim-hipped, white swimmer from an affluent family in Austin, Texas, who began her physical transition only after she'd become an adult and developed in full as male.

It was against this complicated political backdrop that Blackburn and Jackson concluded their remarkable, confusing, and—in an historical sense—devastating exchange:

Senator Blackburn: "Can you provide a definition for the word
    woman?"
Judge Jackson: "I can't."
Senator Blackburn: "You can't?"
Judge Jackson: "Not in this context. I'm not a biologist."
Senator Blackburn: "The meaning of the word *woman* is so unclear
    and controversial that you can't give me a definition?"

Judge Jackson: "If you're asking me about the legal issues related to it—those are topics that are being hotly discussed, as you say, and could come to the court."

Jackson's non-answer left many of us feeling empty, and it opened her up to unwarranted public ridicule. She carried dirty water for both the right and the left that day, which was unfortunate given the moment. She was correct in her final response, of course. This issue is likely to come before the Court. The traditional understanding that most of us have, regardless of political party, that a *woman* is a *female* (and a *man* is a *male*), that what makes someone *female* (or *male*) is their physical form including their reproductive biology, and generally that *sex* is about our biology as male or female, is being challenged in all three branches of government. Because she could be involved in its resolution, the nominee couldn't answer the senator's question. Still, for outsiders to this judicial etiquette, the thread she left hanging was jarring.

The Definition of Sex in Progressive Advocacy

In recent years, progressives who work on sex and gender have coalesced in support of two related goals. Mainly, they want society to accept and the law to treat as equals people who don't conform to gender norms. That's goal one, and it's immediate and practical. Goal two is ideological. It's a society that's blind to sex. This would have been considered futuristic until just a few years ago. But as the exchange between Blackburn and Jackson signaled, it's here and it's concrete. Only certain progressives are committed to this second goal, but the answer to the question "What is sex?" from progressive advocacy is informed by the ideas embedded in both.

Progressives, especially those who care about sex and gender, typically describe themselves as *feminist*. There are factions here too. In very general terms, some have feminism's original goals—empowerment of and equality for *females*—as their central focus and so work on issues like abortion, childcare, wage disparities, glass ceilings, and sexual violence. These are traditional women's issues because they relate to our sex-linked biology and to expectations others have of us because of that biology. In other words, they relate to the Woman Question. *Misogyny* for traditional feminists means diminishing females because of their female biology.

Other feminists have moved on from *sex* to *gender*, including not only to gender roles—which remain the traditional stuff of feminism—but also to gender expression and gender identity. As we saw in the example from the National Women's Law Center just now, and also at the end of chapter 1 in the discussion of Caster Semenya's case, by putting gender identity at the center of their scholarship and advocacy these feminists have turned *misogyny* on its head, redefining it to mean differentiating between females and males who identify as women. It's misogyny, in other words, not to dispense with the biology and to see that we are all women. Old-school feminists are described by feminists of this more recent ilk as "radical" (turning that word on its head too) and as TERFs—a slur in their world that stands for "trans exclusionary radical feminists." In the words of one Twitter warrior, because I'm not blind to sex—regardless of my views about transgender people otherwise—I'm a "massive TERF."

Some institutions embrace both feminisms. As I write, the ACLU and the United Nations both support two different groups within their organizations, a women's rights project and a gender or LGBTQ rights project. These often work in tandem but occasionally sparks do fly, as they did in 2022 when the ACLU's LGBTQ project tweeted out a Ruth Bader Ginsburg quote on pregnancy and abortion that changed Ginsburg's word *women* to *people*. (In the aftermath of this debacle, Michelle Goldberg of the *New York Times* commented, "'Some people oppress other people on the basis of their reproduction' is just not really an accurate way, I think, of describing centuries of patriarchy.") But other institutions are convinced either that the two sets of issues are so substantially the same that they belong under the same umbrella or else that *gender* is a more important focus than *sex*.

This second approach is evident in the transformation of university "women's studies" departments into "gender studies" departments, which have become places where any discussion of the female body as such is anathema. As I write, the ACLU has apparently solved its awkward coordination problem by retaining a single communications person for the two groups: a transgender woman who comes out of the trans rights movement. It appears that the movement is keen to control the messaging on the progressive side of the abortion debate because it sees the development of bodily autonomy law in matters relating to reproductive biology as key to its interests in access to gender-affirming care. In any event, the question who is a "good" or "bad" feminist in this period generally depends on which faction you're in and who's asking.

Most feminists who consider themselves liberal or progressive—and I count myself among them—subscribe to the first, practical goal. We believe that society should accept and the law should treat as equals people who have historically been marginalized, often to the point of invisibility, on the basis of sex- and gender-based differences. This includes people with disorders of sex development (DSD); people who are gay, bisexual, transgender, and gender queer; and people who don't necessarily have a deep-seated inner sense of themselves as any of these things but simply reject gender norms.

The words we use to describe each of these groups today may not have existed in the past and if things continue as they've been going they'll keep morphing, but sex- and gender-diverse people themselves have always existed. The question isn't whether they're "real" but rather how a society and its laws treat them. From common sense and experience, we know that societies do a lot of damage when they adhere to strict binary gender norms and cast down, criminalize, or otherwise marginalize people simply because they're different. If this isn't intuitive for you, just think about something that's central to who you are—your religion, for example, or a disability—that it would hurt to have to hide or reveal. In many countries and communities around the world, being out as gay or trans still amounts to a death sentence. This is unquestionably wrong. The immediate priority for advocates working in this space is therefore to have us see gender diversity as natural and at least neutral. As one of my transgender friends puts it, "No more remarkable or consequential than being a ginger."

Beyond societal acceptance, as we saw in the introduction, advocates for this first practical goal also prioritize securing basic civil rights for gender-diverse people. These are the rights most of us take for granted and for which there's no rational justification for different treatment. Getting seated at a restaurant. Checking into a hotel. Renting an apartment. Signing kids up as patients in a pediatric practice. Getting and keeping a job, including as a teacher, and so on.

Those whose focus is this first practical goal include many transgender people who don't deny that sex is binary. They acknowledge that their sex is one or the other (for example, female) and they want the one they don't have (male) because that one (the male one) best aligns with their inner sense of themselves (as male). They want access to male hormones and maybe to male gender-affirming surgery so that they can match their physical phenotype to their gender identity. They are male "inside" and to the extent possible they want also to be male "outside." They aren't working to take down sex and gender classifications, or single-sex spaces, or gendered artifacts; they affirmatively support them. A transgender boy wants to use the boys' bathroom, not a gender-neutral or the girls' bathroom. Sometimes he may want to play on a boys' sports team, and so on. Their advocates want these things for them too. In litigation, they don't argue that sex and gender are both social constructions that need to be brought down; they argue that they should be preserved but with a door that trans people can come through.

Those whose goal isn't merely practical but also ideological see *transcending* sex as an inherent good without regard to its effects on equality or on other goals and values like safety, privacy, and public health policy. They might want to be able to take puberty blockers forever so that they don't experience either puberty and thus continue to appear as they did in late childhood, as either/or. If they've experienced some part of puberty, they might want to microdose with cross-sex hormones to acquire more of an androgynous phenotype. They might describe themselves as "nonbinary" or "gender queer" rather than as "transgender." In contrast with their practical counterparts, they do advocate for taking down sex and gender classifications, spaces, and language. In their utopia, clothing is "unisex" or "genderless." The wall between men's and women's restrooms comes down and the words on the door are simply "Restroom" or "W.C." Sex and gender are both taken off of birth certificates, driver's licenses, passports, and applications of various sorts. *Latina* and *Latino* become *Latinx*. If we have to have pronouns, it's not *he* or *she*, it's only *they*. They may or may not acknowledge that value is lost along the way, but where they do, it's worth it.

It's worth it, they say, because the focus on our sexed bodies and the classification of individuals according to that biology "essentializes" us— reducing us to just that fraction of who we are. Moreover, sex and gender are both like a physical box whose strictures are reinforced by toxic social norms around masculinity and femininity, none of which are conducive to individual freedom. If you happen to believe that you embrace your body and these norms because you embrace nature or as an expression of autonomy—as part of curating *your* authentic self—you're either ignorant or deluded, mesmerized by messages designed to keep you constrained. Finally, all of this mesmerizing is especially bad for females (whatever their gender identity) who are one with gender-diverse people (whatever their sex) because, just like living in the "wrong" body can be debilitating for a trans person, living in a female body—built as it is for gestation, breastfeeding, and childrearing—is enslaving for females. As Victoria Smith—a feminist writer who disagrees with this perspective—condenses the progressive argument for sex blindness, it's that, "The best feminism (like the best definition of 'woman') dispenses with the dead weight of female biology."

From this vantage point, the real answer to the question "What is a *woman?*" is something other than female biology. The real answer to the question "What is *sex?*" is something other than "our natures as male or female."

The Strategy to Change the Definition of Sex

However they're ultimately motivated, to take down the traditional biological definition of the word *sex*, progressive advocates working in this space have focused on two related efforts. The first is exporting the idea from academia that sex as we know it—as a complete body designed toward reproductive ends—isn't real but rather socially constructed in service of the patriarchy. The second is changing the common usage and legal definitions of the words *male* (and *man* and *boy*) and *female* (and *woman* and *girl*) so that they're consistent with this deconstruction. Here's a bit more detail on each of these two efforts.

Sex (as We Know It) Isn't Real

The argument that sex is just a social construction is grounded in certain scientific and medical facts about primary and secondary sex characteristics. These *facts* are then used to make *arguments* about the "true" nature of sex. Because it's important to see the difference between facts and arguments, I've made a point of italicizing these two words.

Advocates for the proposition that sex isn't real use the *fact* that some secondary sex characteristics like height sort bimodally—in other words, although men are taller on average, some males are short and some females are tall—to make the *argument* that physical masculinity and femininity is a false dichotomy. Indeed, traits like height are not sex characteristics at all, the *argument* goes: when we think we see a male or a female based on their physical form, we're operating on the basis of this false dichotomy. Characterizing these sex differences as "myth," they then go on to illuminate us on the "facts" that, by any rigorous analysis, are just bad *arguments* given the key points they omit.

For example, in response to the question "What's really happening with transgender people's participation in sports? Aren't there biological advantages?" the Human Rights Campaign (HRC) ignores—implausibly, given the context—the *facts* of testicles and testosterone-driven puberty when it answers with the *argument*: "The reality is that all female athletes—transgender and non-transgender—have different shapes and sizes, have different strengths and weaknesses." Because of this, the HRC concludes, the former have no "unfair advantage."

Advocates further use the *facts* that secondary sex characteristics like body fat are helpful but not necessary for reproduction, and that they drive a lot of the toxicity around masculine and feminine gender norms, to *argue* that they're better characterized as "myths and stereotypes" and that *sex* itself is best defined without them. That is, they *argue* that it's best to define *sex* only on the basis of the primary set. They then proceed to deconstruct that set.

As we saw in chapter 1, we've learned a lot over the last century about our primary sex characteristics, including about our chromosomes, gonads, endocrine systems, and genitals. We've learned that in humans, more than 99 percent of the time these characteristics sort in a distinctively binary way. More than 99 percent of us have either testicles and an androgenic endocrine system or ovaries and an estrogenic one. This is why when we call out the sex of a child at birth, we're right more than 99 percent of the time.

These are scientific *facts* we can rely on. Yet advocates of the idea that sex isn't real home in on the *fact* that *all* the characteristics don't always sort this way to support the *argument* that the male-female binary is also a myth and that sex actually exists on a spectrum. For example, they use the *fact* that 1/100,000 males have a DSD called 5-alpha reductase deficiency (5-ARD), which results in the underdevelopment of the external genitals in utero, to *argue* that sex is constructed—that is, "assigned"—not just recorded at birth. Moreover, because it is a *fact* that none of our primary sex characteristics standing alone is responsible for reproduction and all are used for more than just this, the *argument* goes that neither the characteristics nor the set can be defined by their reproductive role.

Specialists have learned that people with DSD tend to identify with their biological sex regardless of the appearance of their external genitals or their sex in rearing. For example, males with 5-ARD who are raised as females are still likely to express a male gender identity. Based on these *facts*, the *argument* is that brain sex, which is related to gender identity in at least some people, exists as an innate and unchangeable primary sex characteristic. In other words, they add gender identity to the primary set. That not everyone has a gender identity— "none" is among the options—and that it's not a factor in reproduction is no more relevant, the *argument* goes, than it is that someone has or doesn't have a uterus.

Finally, if sex was ever real it's not anymore because most of our sex characteristics are now alterable using drugs and surgical procedures. By taking birth control pills year-round, I can avoid ever having a menstrual cycle and thus this multifaceted aspect of being female. A male child can take blockers at the onset of puberty and estrogen thereafter to avoid physical masculinization and thus this multifaceted aspect of being male. There's also gender-affirming "top" and "bottom" surgery—breast augmentation, removal of the penis and testicles, and construction of a neovagina. People of either sex can "microdose" with cross-sex hormones and have facial reconstruction surgery to curate an androgynous form. Given that all of this is now possible, the *argument* is that only our chromosomes and our gender identities are actually fixed or immutable. To the extent sex is real, this is it.

Then comes the coup de grace: for policy purposes, including for purposes of defining *sex*, of the two remaining sex characteristics—chromosomes and gender identity—the latter is most important.

If you want a taste of this approach to deconstructing and then redefining sex so that it's (only) gender identity, read Jennifer Finney Boylan's 2023 op-ed in the *Washington Post* called "To Understand Biological Sex, Look at the Brain, Not the Body." In that piece, Boylan goes straight for rare chromosomal anomalies, then turns to females who have had hysterectomies before concluding that for deciding who is a woman, "It might be that what's in your pants is less important than what's between your ears."

Another great example is U.S. appellate court Judge Jill Pryor's dissent in *Adams v. School Board of St. Johns County, Florida* (2022)—the case I referenced in chapter 1 involving a school board's decision to restrict bathrooms to students based on their sex at birth. Pryor began by disagreeing with the majority's use of the standard definition of *sex* as sex at birth on the grounds that it's insufficiently focused on specific primary sex characteristics. She then added "neurological sex and related gender identity" to the primary set and found that, like chromosomes, these are "immutable" where the rest aren't. She concluded that "when there is a divergence" between chromosomes on the one hand and "neurological sex and related gender identity" on the other, the latter "are the most important factors for determining sex." Bottom line for Pryor: *sex is gender identity.*

For what it's worth, I'm on the record as disagreeing with school boards that would keep trans kids out of the bathrooms that align with their gender identity. But you don't need to take down sex to get to that policy result. Given that it's one of the hotbeds of conservative sentiment around sex and gender, it's especially not good political strategy to do this in Florida.

All of these arguments depend on significant erasures: the erasure of reproduction as part of the definition of *sex*; the erasure of the whole body in favor of a focus on the bits; and the erasure of norms, how people almost always are and how their traits almost always distribute. They also depend on the assumption that sex as we know it is absolute, that it doesn't already contemplate genetic mutations, developmental differences and disorders, and diseases and conditions that affect our physical form and function. Finally, they depend on an ideological preference for characterizing sex as a social construct, a stereotype, and a myth—based on the view that sex isn't good or shouldn't matter.

Changing Common Usage and the Legal Definitions of the Operative Words

Changing the meaning and definition of the operative words so that they're aligned with this political project is key to progressive strategy not only because lawmakers use definitions to make decisions but also because, as cognitive scientist Lera Boroditsky explains, language "shape[s] the way we think," including "how we create knowledge and construct reality." When a language has more words for the color blue, people know and see *blue* in a more complex way than do those that have fewer. Learning of a Russian *invasion* of Ukraine and the *war* that resulted constructs our understanding of what happened in a different way than does a broadcast that tells us that Russia is conducting a *special military operation* there.

Again from Boroditsky, "Changing how people talk" over time "changes how they think." It's because of this that people in positions of power have long included the tool that is manipulating language—the words we know, how they're defined, and those we're not allowed to speak—in their policy toolbox. The progressive ask in this context is not just that we dissociate the words *woman, female, pregnancy,* and *mother* from the female sex, and the words *man, male,* and *father* from the male sex, but that we divorce sex from our natures entirely. Indeed, as we saw in chapter 1, in some progressive circles the term *biological sex* itself is considered transphobic because it brings us back to what they want us to unsee.

The advocacy work here includes periodically updating the definition of the word *transgender*, arguably with a view to its ultimately encompassing all of us. In the last decade, the word has gone from a noun to an adjective—using it as a noun is now verboten—and the adjective from describing someone who has had a sex change operation to someone who has used cross-sex hormones to someone who has lived cross-sex to someone who identifies themselves as cross-sex without the requirements of time or medicalization to—in the words of the Biden administration in 2022—"anyone whose gender identity or gender expression is different from their sex assigned at birth."

This last definition makes *transgender* an umbrella term that includes people who reject gender norms without being motivated by an internal sense of themselves as sexed or gendered. In other words, although I am female, if I don't also have a deeply held inner sense of myself as such or I don't at least perform femininity, I am trans. If you are male and don't perform masculinity or have a corresponding inner sense of yourself as such—maybe because it's never occurred to you to think about this one way or the other and so you'd say "none" if you were asked—you are trans. Given that both masculinity and femininity are simultaneously described as "toxic," you see where this is going.

Each of these redefinitions automatically increased the numbers of people who can be described as trans. They even allow us to identify people posthumously as trans. In a *New York Times* op-ed, the novelist Peyton Thomas did this to Louisa ("Lou") May Alcott, whose papers show that she—Thomas suggested Alcott's preferred pronoun today would be *they*—eschewed the dress codes and conventional strictures associated with being female in favor of the different dress codes and freedoms associated with being male. Thomas's op-ed together with the responses are a terrific illustration of our different feminisms.

The advocates' two-pronged strategy—arguing first that *sex* as we know it isn't real and then that the definition of the words should be de- or unsexed—was evident in the 2019 and 2021 versions of the Equality Act, the legislation also known as H.R. 5 that I began to describe in the introduction. It's designed to extend the protections of the 1964 Civil Rights Act to sexual orientation and gender identity.

The 1964 act prohibits discrimination on the basis of "race, color, religion, sex or national origin." The earlier 2015 and 2017 versions of the Equality Act sought to prohibit discrimination on the basis of sexual orientation and gender identity by *adding* these to the list:

Section 201 of the Civil Rights Act of 1964 (42 U.S.C. 2000a) is amended—. . . by inserting "sex, sexual orientation, gender identity," before "or national origin."

But the later 2019 and 2021 versions opened up *sex* itself, expressly redefining it to include sexual orientation and gender identity:

> Section 201 of the Civil Rights Act of 1964 (42 U.S.C. 2000a) is amended . . . by inserting "sex (including sexual orientation and gender identity)," before "or national origin."

The bill as revised erases the category differences between *sex, sexual orientation,* and *gender identity.* Lest there be any confusion about the lawmaker's intent, the new phrase, "sex (including sexual orientation and gender identity)," is then used throughout the legislation.

A definitions section expands on this revolutionary erasure by explaining that *sex* doesn't only "include" *sexual orientation* and *gender identity* but also *sex stereotype* and *sex characteristics, including intersex traits.* It defines *gender identity* consistent with the Biden administration's broad 2022 definition of *transgender* as "the gender-related identity, appearance, mannerisms, or other gender-related characteristics of an individual, regardless of the individual's designated sex at birth." *Sex stereotype* isn't further defined but it's already a legal term of art meaning either an accurate generalization about males or females— (for example that males are generally taller than females), or a false one (for example that females can't do math). The legislation doesn't further define *sex characteristics or intersex traits* either and so we don't know whether they're intended to mean our primary and secondary sex characteristics, or just the primary set.

As I noted in the introduction, most glaring is that the proposed redefinition of *sex* omits the traditional biological definition and common understanding of the word as—per the Oxford Dictionaries—"either of the two main categories (male and female) into which humans and most other living things are divided on the basis of their reproductive functions." To the extent this meaning remains, it's only implicit—through the phrase "'sex' includes" and in its deconstructed form as "sex charac-teristics, including intersex traits."

The movement's talking points tell us that redefining *sex* and then prohibiting—as the Equality Act would further do—almost all sex classi-fications is good for everyone. In particular they insist that these changes would be good for women—not only for those in the LGBTQ communities but all women. In support of this claim, they point to the redefinition of *sex* to include *pregnancy, childbirth, and other related medical conditions*, except that these are already covered by existing law. They offer that the legislation would add protections for breastfeeding, but beyond "and other medical conditions"—a catch-all that courts have said doesn't include breastfeed-ing—the text doesn't mention it. Finally, they claim that it would be good for all women if the movement mantra "trans women *are* women" were codified through the erasure of sex from law. Especially as they see female biology and hard-won rights being erased, not added, women who aren't trans may not understand the connection to their welfare.

At the Equality Act hearing in 2019, the National Women's Law Center's emissary—who was there in support of the legislation as part of the organization's decision in that period to focus on trans advocacy and in the process to articulate the relationship to its traditional mission—gave this stunningly patronizing and tautological response to those of us who were concerned about that connection. I quoted part of it already in the introduction, but here it is in full:

> Because transgender women are women, and so all of us need this protection together . . . why would the National Women's Law Cen-ter and the host of women's rights organizations be here in support of the Equality Act if it was going to harm women? We are the ex-perts on this. This is what we do, day in and day out, across sectors, workplace, you know, healthcare. Workplace, justice, education, all

of these areas, this is what we do is we fight for women's rights. And so please look to us as the experts on whether or not this bill is good for women and LGBTQ people.

A mere seven months after Ketanji Brown Jackson's confirmation hearings, in October 2022, the editors of the *Cambridge English Dictionary* decided that the word *woman* now means *both* "an adult female human being" *and* "an adult who lives and identifies as female though they may have been said to have a different sex at birth." Although it was not yet fully synonymous, Cambridge's definition of the word *female* was also well on its way to becoming anyone—including a male—who identifies as female. Its first definition of *female* remained "belonging or relating to women" but as an illustration of word usage, it now included "She was the school's first trans female athlete." Its second definition and illustration were intact, as "belonging or relating to the sex that can give birth to young or produce eggs" with the illustration "Female lions do not have manes." They don't have testicles either, but many trans female athletes do. Finally, although the Cambridge dictionary continued formally to define *sex* as "the physical state of being either male, female, or intersex," it also used the word in its definitions of *woman* and *man* to mean something that exists not as a physical state but as something we are "said to be."

Because of the way the editors of dictionaries do their updating, it's likely that Senator Blackburn's question to then-judge Jackson contributed to this outcome. That is, contrary to the senator's intention to protect a traditional definition of *woman*, the Internet searches her question generated were likely a substantial cause of the spike that changed it. As Cambridge spokesperson Sophie White explained to the *Washington Post*, "The dictionaries are compiled by analyzing more than 2 billion words. . . . We regularly update our dictionary to reflect changes in how English is used, based on analysis of data from this corpus." She added that its editors "carefully studied usage patterns of the word *woman* and concluded that this definition is one that learners of English should be aware of to support their understanding of how the language is used."

Dictionary.com made *woman* its Word of the Year in 2022. It explained in the announcement that "searches for the word *woman*" on its platform "spiked significantly multiple times in relation to separate high-profile events, including the moment when a question about the very definition of the word was posed on the national stage." By far the "biggest search spike" was "during a confirmation hearing for Judge Ketanji Brown Jackson, who in April became the first Black woman to be confirmed as a U.S. Supreme Court justice. Specifically, the surge in lookups came after she was asked by Senator Marsha Blackburn to provide a definition for the word *woman*." The announcement went on to explain:

> It was a rare case of not just a word in the spotlight, but a defini-
> tion. We at Dictionary.com weren't the only ones to take notice.
> The prominence of the question and the attention it received dem-
> onstrate how issues of transgender identity and rights are now fre-
> quently at the forefront of our national discourse.

This technical explanation drew criticism from people who saw the not-so-invisible hand of progressive advocacy at work. Indeed, the trans movement's fingerprints were all over the concluding lines of Diction-ary.com's explanation for its choice of Word of the Year:

> More than ever, we are all faced with questions about who gets
> to identify as a woman (or a man, or neither). The policies that

these questions inform transcend the importance of any dictionary definition—they directly impact people's lives.

We also see the movement's fingerprints in Cambridge's description of a trans person as someone whose sex at birth is "said to be" other than it is; in its decision to include trans women and girls not only in the definition of *woman* but also in the definition of *female*; and in its use of trans girls in sports as illustrative of *females*. Still, it's also true that Senator Blackburn's decision to ask Judge Jackson the question "Can you provide a definition for the word *woman*?" contributed to the word count, which changed usage, which changed dictionary definitions, which—because of the Supreme Court's preference for the textual interpretation of words—has the potential to change legal meaning. Depending on how this turns out, one might even draw from sports and call the senator's decision an "own goal."

The extraordinary success of the redefinition project was most evident to me two months later when, on December 22, 2022, just three weeks after its beta release, one of my sons and I were playing with ChatGPT for the first time. ChatGPT "learns" the answers to questions by "scraping" or "chewing" portions of the Internet. We asked it, "Is a transgender woman a 'woman'?" Its automatic response went a step beyond Cambridge. ChatGPT didn't just add certain trans women to the category, which was otherwise still composed of adult human females; it de-linked women from the female sex altogether. "Yes," it said, "trans women are women":

> Being a woman is not determined by one's anatomy or biology at birth, but rather by one's gender identity. A person's gender identity is their inner sense of being male, female, or something else. A trans woman is a person who identifies as a woman and may take steps to transition to living as a woman, which may include taking

hormones, undergoing surgery, or changing their name and pronouns. It is important to recognize and respect each person's gender identity and to use the name and pronouns that they prefer.

According to ChatGPT at the end of 2022, females aren't women if they don't have an inner sense of themselves as gendered, but males are women if they think of themselves as such—even as they may continue to live and present as men. Because of how they "learn," chatbots aren't always right and their answers will evolve over time. Still, as we go they'll be extraordinarily powerful—like dictionaries on steroids.

The Woman Question has taken many different forms across the millennia but it has never before been decoupled from our physical nature, which has always been the through line. It has never before been entirely about gender. How we come out of this contested moment doesn't, I think, depend on whether some trans women are included in an expanded understanding of *woman* for ethical, political, or even practical reasons, but rather whether sex as a characteristic (regardless of one's gender identity) continues to matter in law and policy. This will be resolved based on whether we can make a good case that it still matters to our lives. If we do this, because of how language works, we should have the necessary words. In the meantime, savvy public speakers continue to convey relevant meaning.

In her remarks on the White House lawn after she took the oath of office on April 8, 2022, Justice Jackson thanked everyone who had supported her along her journey, including "each member of the Senate," whom she noted, "fulfilled the important constitutional role of providing advice and consent." Then she returned to American history, which she made that day in one more way, by reaching the highest rank ever in the U.S. government for a female descendant of American slaves.

The notes that I've received from children are . . . especially meaningful because, more than anything, they speak directly to the hope and promise of America. It has taken 232 years and 115 prior appointments for a Black woman to be selected to serve on the Supreme Court of the United States. But we've made it. We've made it, all of us. All of us. And our children are telling me that they see now, more than ever, that, here in America, anything is possible.

They also tell me that I'm a role model, which I take both as an opportunity and as a huge responsibility. I am feeling up to the task, primarily because I know that I am not alone. I am standing on the shoulders of my own role models, generations of Americans who never had anything close to this kind of opportunity but who got up every day and went to work believing in the promise of America, showing others through their determination and, yes, their perseverance that good—good things can be done in this great country—from my grandparents on both sides who had only a grade-school education but instilled in my parents the importance of learning, to my parents who went to racially segregated schools growing up and were the first in their families to have the chance to go to college.

I am also ever buoyed by the leadership of generations past who helped to light the way: Dr. Martin Luther King Jr., Justice Thurgood Marshall, and my personal heroine, Judge Constance Baker Motley. They, and so many others, did the heavy lifting that made this day possible. And for all of the talk of this historic nomination and now confirmation, I think of them as the true pathbreakers. I am just the very lucky first inheritor of the dream of liberty and justice for all.

To be sure, I have worked hard to get to this point in my career, and I have now achieved something far beyond anything my grandparents could've possibly ever imagined. But no one does this on their own. The path was cleared for me so that I might rise to this occasion. And in the poetic words of Dr. Maya Angelou, I do so now, while "bringing the gifts . . . my ancestors gave." I—"I am the dream and the hope of the slave."

Part II

Sex
Matters

4

Sex Is Good!

At THE GALLERIA DELL'ACCADEMIA in Florence, Italy, down a long corridor in a two-story rotunda, is the *David*, one of the world's most famous sculptures. Because you go there to see this work of Michelangelo's, you'll arrive thinking you're prepared and that you won't be stunned, but then you are. Utterly. The tourist guide's refrain, and it's true, is that the *David* doesn't disappoint. It's not just its size, at over thirteen feet, which was conceived to impress; or its display under a glass dome in natural light after the artificial illumination of the hallway; or even the artist's extraordinary talent, which is indescribable—though much ink (including mine) has been spilled trying. It's that the *David* is extraordinarily beautiful both as a work of art and as a man. His nakedness—in the nude sense of the word—is part of what art historian John Paoletti explained "still causes people to spend so much time in front of him, whether they would admit this or not."

In 1501, when Michelangelo was just twenty-six, he was commissioned to carve an Old Testament figure out of a difficult piece of marble. He chose David as his subject, but not the famed king, lover, or father. Rather, he chose to depict the young warrior whose inner strength of mind and outer strength of body combined to defeat Goliath. Michelangelo further chose to portray David as the unknown man in the moment before he throws the stone, not as the eternal hero in the aftermath of his victory alongside the head of his enemy, as had been the artistic tradition.

In that earlier moment, David's head is turned and his eyes are focused on his target. There is tension in his face beyond the eyes as his deep brow ridge is furrowed and his nostrils are slightly flared. But the rest of his perfectly proportioned form—equidistant from head to hips and hips to feet—is still loose, ready but not yet fixed. The sprinter before the start or the striker before the penalty. His locks remain in place; his right arm is by his side and his left leg is bent, waiting. But even without the physical strain that inevitably follows, we see the cuts and the promise. His traps, deltoids, pectorals, biceps, and triceps flow into his abdominals, hip flexors, and glutes, and then into his quads, hamstrings, and calf muscles. All banded together by tendons and ligaments, also in relief. This would be an implausible display absent movement but he is lean, and it was Michelangelo, and so it's there.

It's been said that the *David*'s head is too big and his genitals are too small—out of proportion and inharmonious relative to the rest—because the artist sought to convey that the young David had cultivated his mind and tamed his nature. Still. Michelangelo's contemporary the Renaissance painter and writer Giorgio Vasari waxed poetic about "the most beautiful contours of his legs" and pronounced, "[O]f course those who see this need not care to see other sculptures made in our times, or in other times by any other artist."

It can be easy these days to forget that sex—the male and female body—is good; and not just because some trans rights organizations are working to convince us that it's not or best ignored. There are plenty of feminists who think that freedom and equality for females can be achieved only if we reject what they see as the dead weight of female biology. And there are innumerable women, like me, who love and are perfectly comfortable in their bodies but who want the same opportunity as men to be educated and to work during their prime reproductive years and so welcome the drugs and procedures that allow us to time shift and minimize the harder effects of our sex.

The pervasive negativity about sex extends to boys and men. It's common these days to hear natural male traits from men's sexuality to boy energy and lots in between described as bad—things to be shamed or medicalized—without distinction between normal and toxic masculinity. The news frequently features stories about sexual violence and sex discrimination and their disproportionate impact on females, as well as the disproportionate impact of globalization, organized violence, and drugs on males. We need to be familiar with these troubling facts if we want to do something about them, but they're undoubtedly part of the constant drumbeat telling us that sex is a problem. It's a rare day that we hear the French *Vive la différence!* or any reflection on Ruth Bader Ginsburg's insistence that "'inherent differences' between men and women . . . are cause for celebration[.]"

But our inherent natural differences *are* a personal and collective good. By *good* here I mean valuable, something that has clear positive weight for us and our societies. Especially in this age of naysaying and deconstruction, it's important to see that, to remember why, and to say it out loud, at least so that when we're working on the costs we don't throw the baby out with the bathwater. I'm going to argue that sex is good for three distinct reasons. Sex is good for procreation, it's good for sexual pleasure, and it's good for its aesthetic value.

Sex Is Good for Procreation and Procreation Is Good

I'll start with the most obvious. Sex—the male and female body—is necessary for procreation and procreation—having and raising children—is good. It's good for the species, it's good for our societies, and it's good for each of us individually.

Procreation is good for the species because without it the human race wouldn't survive. It's why we exist, why we're born male or female, and why our bodies function as they do. Whatever else we might do with them, however else we justify our own individual existence, this comes first. The survival of the species isn't typically what motivates us to have kids, nor does it motivate policymakers who are more narrowly focused on their particular geography and political goals, but the idea is that in the aggregate, enough of us will contribute for our own reasons so that the imperative is met. If most of us produce at least an heir, if not also a spare, our communities will be sustained, and their sustenance will in turn sustain the species.

Procreation is also good for our societies—our towns, states, regions, and countries—and for the very same reason. Without both males and females, it's impossible to sustain a community, and without both more or less in balance and interested in having kids, it's hard to sustain a healthy society. A fabulous graphic by statistician Kaj Tallungs speaks to these basics. It reminds us that sex isn't only a coin toss as to any given birth, it's a coin toss for the whole population. In this country and others, the pattern is essentially the same: the two sexes mirror one another statistically, with a small natural surplus of males early in life, a bigger surplus of females later in life, and parallel changes otherwise. The graphic also tells us that the numbers in each generation and the relationship of the generations to one another are important facts for those charged with thinking about the public's welfare.

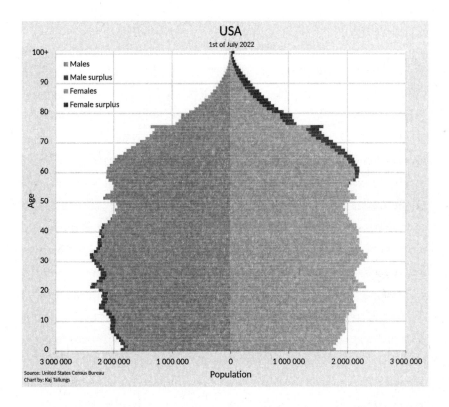

USA

1st of July 2022

■ Males
■ Male surplus
■ Females
■ Female surplus

Source: United States Census Bureau
Chart by: Kaj Tallungs

Population

Governments track birth rates and population numbers in sex and age categories so that they can plan for the military, childrearing, education, health care, pensions, and more. It's true that they haven't always officially monitored sex, but they've always had the information handy when they needed it, through gendered names and religious records. Although gender norms have changed in many parts of the world, the traditional patterns—mostly males in the military and in physically taxing manual labor, mostly females focused on childrearing and other service work—are still prevalent. We can go out of our way to pretend it's otherwise, but as I write, in the United States more than 80 percent of active-duty military personnel are male, over 90 percent of coal miners and construction workers are male, and over 90 percent of childcare workers are female. On average, females also live five years longer.

Information about birth rates, sex, and age is also collected to know the relationship of children to seniors and of both to those of working age—because these relationships are key to economic planning. Such statistics have become a regular news item as women are having fewer children and people are living longer after retirement. The combination is a problem because all societies count on people of working age to take care of their younger and older members—whether directly or indirectly through taxation. Look at the graphic again. The kids who are now in the under-ten age bracket will—in the next ten to twenty years—be contributing to the support of the people who are now fifty-five and older, and there are a lot more in that second bracket. It's no surprise that debates over entitlements are often front and center in our politics.

The trend of women having fewer children will continue around the world, in some places dropping far below *replacement level*, which is usually 2.1 children per female. (Since a man can't bring a baby to term, for a man and a woman to replace themselves at least once over, the woman needs to have two children who survive to reproduce.) This decline in *replacement rate fertility* has governments talking in "crisis" terms. Here are two recent headlines:

POPULATION IN MORE THAN 20 COUNTRIES TO HALVE BY 2100.
Two choice quotes from this 2020 article in Al Jazeera: "More than 20 countries, including Italy, Japan, Poland, Portugal, South Korea, Spain and Thailand, will see their numbers diminish by at least half by the year 2100," and "These forecasts suggest good news for the environment, with less stress on food production systems and lower carbon emissions, as well as significant economic opportunity for parts of sub-Saharan Africa. . . . However, most countries outside of Africa will see shrinking workforces and inverting population pyramids, which will have profound negative consequences for the economy."

CHINA'S POPULATION FALLS, HERALDING A DEMOGRAPHIC CRISIS. The article in the *New York Times* in 2023 elaborates: "[F]acing a population decline, coupled with a long-running rise in life expectancy, the country is being thrust into a demographic crisis that will have consequences not just for China and its economy but for the world."

The United States isn't being spared. Our fertility rate has dropped below 1.8 births per female, in all racial and ethnic groups. As Lyman Stone of the Institute for Family Studies explained,

> The loss of babies in the U.S. will have momentous consequences in the future, as the next generation is smaller than the one before it. The economic consequences of low fertility can be dire, as can other outcomes policymakers care about, ranging from military recruitment to instability in electoral coalitions. But these societal effects are of secondary importance to what low fertility means in the lives of those who experience "missing births," ranging from rising loneliness to aging alone to less happiness. The consequences of low fertility today will echo through Americans' increasingly empty homes for decades to come, leaving millions more people isolated and adrift from wider society as they age.

Because women having children at replacement rate is such a big deal for our societies, people are studying why the numbers are dropping. There's some debate about whether sperm quality (and thus male fertility) is declining, but the focus is on declining female fertility. The working explanation is that when they can, women are making economic and lifestyle choices that don't include children, and certainly not as many as in the past. Some are also choosing to delay pregnancy past their *prime reproductive years* in pursuit of the same educational and early career opportunities that used to be available only to boys and men. Although we're fertile from menarche in our early to midteens to menopause in our late forties to early fifties, it's easiest to sustain a pregnancy from our late teens to late twenties—precisely the period during which it can be most educationally and occupationally disadvantageous to have children.

And so governments are looking for a "fertility fix." In some places this means reducing or prohibiting access to birth control and abortion. In others it means cash for kids or gametes. In 2023, for example, "a hospital . . . in Southwest China, announced that college students—but only those taller than 5-foot-5—who donated their sperm could receive 4,500 yuan, or about $660. Sounding like a collective call to action, the announcement concluded with a slogan in pink font: 'I donate sperm. I am outstanding. I am proud.'" Mostly, though, governments support measures linked indirectly to higher fertility rates, especially those that tend to promote "gender equality, marriage and cohabitation, maternal and social support" and "pro family government programs."

Studies suggest pro-natal policies probably won't make much of a difference as the factors governments can influence don't match up to the reasons many young people aren't having kids. There's also a raging debate about whether we should even want replacement-rate fertility, given the climate crisis. But I'll stop here because my point is simply that procreation is essential for societies and so they care about sex.

Just as procreation is good for the species and our societies, it's good for us individually. By *good* I don't mean it's cost-free or that the costs—economic, educational, emotional, whatever they may be in any given instance—are worth it for everyone. But having and raising a child or children is something that usually weighs heavily on the plus side of our personal decision-making scales. Not only are we naturally built for child-bearing and -rearing but our lives are usually enriched when we have kids. It's why people who struggle to have them don't just give up, making assisted reproduction and surrogacy thriving businesses around the world.

I was in Marrakesh when I was writing this chapter. On the road from the airport I passed a small stone building that doubled as a pharmacy and Centre de Procréation Medicalement Assistée. I wanted to know more about the Islamic tradition and so I did some quick research. Starting from the Quran, which teaches that "[w]ealth and progeny are the allurements of this world," Dr. Hossam Fadel, a fertility specialist, explains that "A common supplication of Muslims is, 'And those who pray, Our Lord, grant unto us spouses and offspring who will be the comfort of our eyes.'" It's because this hope is universal that many legal systems describe having kids as a natural and a fundamental right. Although the United States has a terrible history here, involving genocide of the American Indian population, the decimation of enslaved families, and involuntary sterilization of vulnerable subpopulations in the name of eugenics, *procreative liberty* is today as close to inalienable as any of our individual rights get.

But why are kids good for us? I'll begin with the natural piece of the story. Whatever else we might do with them, our bodies are built to procreate. It starts with our male and female architecture—our sexed anatomy and physiology—which we might be able to ignore were it not for the hormones that begin to rage in adolescence. These tell us when we're fertile and encourage us to have sex; facilitate the ups and downs of the menstrual cycle, pregnancy, breastfeeding, and bonding; and sustain and support our own life cycles. As we've already seen, there's testosterone for male sperm and semen production and for building a protective body. There's estrogen and progesterone for the female menstrual cycle, pregnancy support, and building a body that can gestate a child. There's prolactin so that female breasts grow and make milk to feed newborns. There's oxytocin for everyone, so that we fall in love and bond with our babies. There's cortisol, the fight-or-flight hormone, that surges when we need to protect ourselves and our children. The study of the physiological effects of pregnancy and children has long focused on females. But kids affect men too, as we are now discovering—and far beyond testosterone surges in moments of maximal fertility and cortisol surges in times of perceived danger. At least some men, for example, experience drops in sex hormone levels when their partner is pregnant and a child is born—a hormonal phenomenon that is thought to promote commitment, bonding, and love.

Whatever your gender identity, if you're female and have felt a child growing inside you, had your milk suddenly surge and let down when your newborn makes a sound in another room, experienced their rooting reflex as they bobbed with their mouths on your shoulders and chest in search of your nipple, locked eyes with them for longer than you could imagine doing with anyone while they're breastfeeding, been overwhelmed with fear and even disconnected when things go awry, you've experienced some of the most intense emotions possible. These are all hormonal effects. When my children were young, I described them as primal. I remember thinking, for example, that it was deeply wrong to get on a plane when my toddler was still at home—driving to work was one thing but leaving the ground was abandonment. I also remember the inexplicable joy of the most routine milestones—the first smile, word, step, run, book, friend . . . and the irrational pain of the innumerable small disappointments on the way to adolescence and at least physical independence.

Although fathers have only a subset of these physical experiences—pregnancy and breastfeeding are sui generis—they describe parenthood in much the same way. As do adoptive parents. Not every single person, of course. Some wall themselves off from the intensity. I'm thinking here about females with postpartum depression, males who are raised in cultures where allowing themselves these deepest of emotions is not seen as manly or dignified, and people with body and gender dysphoria for whom pregnancy and breastfeeding can be profoundly negative experiences. They do what they can to shut down these hormonal effects. But if most of us didn't open ourselves up to caring about our kids unconditionally or at least unselfishly, they wouldn't survive. Human offspring are physically immature and dependent for much longer than the offspring of most other species, and so reproduction—raising a child to the age when they can themselves reproduce—is about lots more than just pregnancy.

Beyond the natural part of the explanation, creating and sustaining new life gives unique meaning, purpose, and value to our existence. My cousin Nina Lorez Collins has developed companies to support women going through menopause and aging and written extensively about her life as a daughter and a mother. She described wanting to have children at twenty-two this way: "My mother had just died, my father was out of the picture, and like the young black girls in the ghetto who want to have their own child even in the midst of that adversity, I wanted to have a child to love unconditionally and to feel the indelible connectedness to another human being. I wanted to create family." As she recalled it to me, her decision was "almost selfish." But even so, there are four beautiful children to whom she gave the gift of life.

Having children is also a way to build and sustain family lines, relationships, and traditions. In many settings, the birth of a child is a community-wide gift that spreads joy and reinforces membership, belonging, and even—as one of my queer male friends put it—legitimacy. Religious communities, for example, are often child-centered.

A conservative Christian answered my "Why children?" inquiry in a way consonant with the Quranic sentiment described a moment ago: Christians, he said, "believe that humans are the most marvelous thing in creation—in part because they are the only thing in creation capable of marveling." This means that "the raising and formation of the next generation is the most important, complex, and rewarding effort and art in life. . . . All else you might consider investing yourself in cannot be anything but second place." He added that because "Christians are taught to aim higher than happiness—for joy," it doesn't matter that "there are times when one would be 'happier' without the responsibility," since "we so often find joy in our children, and how often do we find joy anywhere else?"

The coolest "Why kids?" conversation I had was with Raffi Grinberg. Raffi and I met in 2022 when I joined Dialog, an organization that brings people together for nonpartisan conversations about important social issues. He'd earlier designed and taught a college course called Adulting 101, in which he encouraged young people to be intentional about planning their lives, including having children. He describes himself as a pro-natalist; he's committed for both moral and personal reasons to having children and—in his and his wife Charlotte's case—to having as many as they can. (In case you're wondering, Charlotte is a hospice and palliative care doctor who also writes about her life and work.) This was Raffi's answer:

> Morally, human existence is good and more human existence is better than less of it. Some people live very difficult lives but by far most of us would say that we're glad that we exist, and humans do a lot that's useful, important, and good that would be lost if we didn't exist. Because we can't know which of us will do the important work and innovation and make the important contributions, deciding against as many kids as possible is possibly to miss creating the humans who would make the most difference.

> Personally, even though our kids will be free to choose how they want to live their lives when they're grown, more rather than fewer kids is also a good retirement plan—including a good loneliness retirement plan—in a time when Social Security and pensions aren't going to do the trick and older people are increasingly sad and alone.

Raising kids is of course costly, he added, but for him "the costs pale in comparison to the benefits of having another kid." He ended with his sense—a lasting gem—that the "happiest people tend to have diversified their assets of meaning. Most of the things we choose to do in life that give us meaning end up giving us heartache *and* joy."

Which brings me to the "happiness penalty" parents supposedly suffer in relation to people who are childless. In the United States, people *without* kids report being 12 percent happier than people *with* kids. This is said to be the biggest such gap in the world. Given how high some general idea of happiness is on their lists of ambitions, it's no wonder many of my students are thinking about forgoing the experience of parenthood entirely. What seems to be clear, though, is that this statistic is only right if you define *happiness* as something completely distinct from *meaning*, to signify *carefree* or *independent*. Nonparents in the United States report a much lower sense of *meaning* than parents do. The stats are also right only if you live in a place where parenting is an individual, not a family or community, commitment. In Portugal, for example, where the burden and joy of childrearing is often shared within extended families, the gap runs in the other direction—parents are 8 percent happier than nonparents.

One last point about the human desire for kids. We're naturally and rationally driven to making and caring for them regardless of our sexual orientation and gender identity—even as these circumstances can obviously complicate matters.

A gay man may not want to have sex with a woman and a gay woman may not want to have sex with a man, but the human desire to procreate remains. More than once I've heard the sadness and frustration some gay people feel because the rest of us can "just have sex and there's a baby." To get just a small sense of this, watch the brilliant "Aunt Mommy" episode of the paradigm-shifting comedy *Modern Family*, in which Mitch's sister Claire, having had too much wine, offers to donate an egg so that he can have a biological child with his husband, Cam.

The same goes for a person who is transgender. Whether my sense of my own sex is concordant or discordant doesn't affect whether I want to be a parent, even as pregnancy for a transgender man can be a really difficult emotional and physiological trade-off. As journalist Freddy Mc-Connell so beautifully expressed this in the documentary I mentioned earlier, *Seahorse: The Dad Who Gave Birth*, going off of testosterone and then pregnancy itself felt like a "total loss of myself" to the point where "I just want to close my eyes and be on the other side of this"—a process his mother revealed was "brave, braver than we can imagine." But he wanted that much to get to that other side to know this deepest of commitments. When he got there he was overwhelmed in the same primal way I've described my own experience with each of my sons: "I have a really strong bond with him," McConnell said of his child. "I just know that I want to be near him, all the time . . . and I miss him when he's in bed at night . . . it's crazy . . . they're a little extension of you . . . I think anyone has the potential to feel that way."

Sex Is Good for Pleasure and Pleasure Is Good

Sex is good for procreation, but it's also good because it's the reason we experience sexual pleasure. For some of us sexual pleasure can be good only if it leads to procreation and for others it's good for its own sake. Regardless, as pleasure is good, sexual pleasure is good. I could stop here but because there are contrarians on this point too, I'll spell it out.

Sexual pleasure is the physical and emotional good feelings, including bonding, that we get from being physically intimate with another person. The experience is broken down by medical experts into the stages of the "sexual response cycle." These stages are standard, meaning that it's natural to experience each of them in this sequence because they correspond to the reproductive imperative. Doctors know them because they see patients who are atypical and need help experiencing what is natural and good.

First comes *desire*, which results from a combination of biological and psychological drives. The physical manifestations of desire last from a few minutes to several hours and may include increased heart and breathing rate, increased muscle tension, and blood flow to the genitals—in males an erection and in females vaginal lubrication and swelling vaginal walls. Then comes *arousal*, which "extends to the brink of orgasm" and during which the "changes begun in the first phase get more intense." Last is *orgasm*, "the shortest of the phases," which "generally lasts only a few seconds" and during which "your body releases dopamine, known as 'the feel-good hormone,' and oxytocin, sometimes called 'the love drug.' These hormones increase feelings of happiness and other positive emotions, and they counteract the 'stress hormone,' cortisol." Orgasm is characterized by "[i]nvoluntary muscle contractions" and "a sudden, forceful release of sexual tension. In women, the muscles of the vagina contract and the uterus may also undergo rhythmic contractions. In men, rhythmic contractions of the muscles at the base of the penis result in the ejaculation of semen."

Sexual pleasure depends on our sex—our physical forms as male or female—and our sexual orientation. In the United States, about 90 percent of us identify as heterosexual, meaning that we're mostly or only opposite-sex attracted. As I write, 4–5 percent of us identify as bisexual, meaning we're attracted to both sexes, and about 3 percent of us identify as homosexual, meaning we're mostly or only same-sex attracted. These percentages change over time depending on how acceptable it is to self-report as something other than heterosexual. Regardless of which group describes you, as the British philosopher Kathleen Stock says, "You like sex with the sex(es) you like, and it seems to start quite early on in life." We sometimes have "voluntary and even pleasurable sexual experiences at variance with it," but ultimately, "sexual orientation is for life, not just for Christmas parties." Stock—author of the bestselling book *Material Girls: Why Reality Matters for Feminism* (2021)—is herself a lesbian.

Starting a few years ago, a small but vocal group of transgender women provoked a firestorm with their claim that sexual orientation is transphobic. *Transphobia* has multiple meanings. The basic one is a fear or hatred of people who are transgender. The more controversial one is the refusal to see trans people as being the sex they say they are; or, put differently, the refusal to ignore their biological sex. The argument is that someone who isn't afraid of and doesn't hate trans people is still transphobic if they see or consider a trans person's sex.

The most extreme version of this claim disallows consideration of a trans person's sex even in the selection of sexual partners. Caring about sex is always unfair toward trans people, the argument goes, and embracing diversity and inclusion is the moral things to do. Doing what's best for the least well off—in this case trans women—is also the equitable thing to do. Veronica Ivy, formerly Rachel McKinnon, is a Canadian American philosopher who is herself a transgender woman. In 2019, she posted this on Twitter:

> I actually think any sexual orientation other than pan is immoral because sexual genital preference is [*sic*] immoral.
>
> You: "I like dick."
> Girl with dick says, "Hey, wanna date?"
> "Oh . . . no . . . I only like dick on guys"
>
> Guy responds to date ad: "Sup girl" . . . guy has vagina
> "Oh, sorry, I only like guys with dicks"
> Both cases trans people are left in the cold. "Genital preferences" are transphobic.
>
> Date genders, and just learn to have fun with their sexy fun time bits, whatever they happen to be.

There's another version of this claim that only disallows consideration of trans people's sex. It seems to have originated with a subset of trans women who, because they identify as women and are primarily attracted to females, call themselves "lesbians." To be clear, if they hadn't come out as trans they would be straight men. They insist that (female) lesbians who don't want to have sex with them are TERFs and (yes) also misogynists. According to them, a good lesbian wouldn't see their (male) sex; she would see only their (female) gender identity. They refer to a "cotton ceiling"—analogous to the "glass ceiling" that blocks women's path to the highest jobs in business—that distinguishes women-born-male from women-born-female and blocks their path to their preferred sexual partners. They've used social media to bully resistant lesbians, and some of their allied organizations have apparently counseled young (female) lesbians against their discriminatory ways.

I've already quoted the philosopher Kathleen Stock on sexual orientation. She's been bullied, deplatformed, and protested for holding supposedly transphobic views to the point that she resigned from her university because it said it could no longer guarantee her physical safety on campus. Her crime? She holds the position, described and rationalized in detail in *Material Girls*, that trans women aren't women because the word *woman* standing alone means adult human female. A piece of this is her rejection of the claim that trans women who prefer to have sex with females are "lesbians." For Stock, lesbians are, by definition, females who like sex with other females. Those who try to shut her down say they're justified because she causes harm to transgender people. She insists that their demands—which are not merely rhetorical—cause harm to females, including to (female) lesbians.

Even when it's not backed up by threats of physical violence and gaslighting, the claim that sexual orientation is transphobic is dangerous for feeding the false stereotype that trans women are men with predatory tendencies, a view that well-informed, caring people, including in other parts of the trans and allied communities, are working nonstop to correct. Most transgender women don't deny the reality of anyone's sex including their own; they know that everyone's sexual orientation—including their own—is sex-based; and they're committed to everyone's bodily and personal autonomy. They're just like the rest of us in these undoubtedly important and ultimately personal respects.

The kindest response to the claim that sexual orientation is transphobic is that it misunderstands or misrepresents the trait, including that it's inherent, not constructed, and instinctive, not rational—and that these attributes are precisely part of its power. The notion that good, moral people should not be free to exclude in the context of our most intimate choices and relationships is misguided. The claim also misunderstands or misrepresents that sexual orientation isn't reducible either to genital preference or to gender identity. In this the claimants make the same mistake as gender theorists who see sex myopically, in disconnected bits rather than in the body as a whole.

It's hard to be kind, though, because the stance is deeply coercive of the last thing that should be coerced: sexual relations. Scholar-activists, frustrated folks with a megaphone, and affiliated social justice organizations have no more business telling us who we should and shouldn't want as a sexual partner than do orthodox religious groups. The further implication that our sex-based attractions ought to be civilized along with other bad traits and ideas is just as damaging as saying they should be civilized according to conservative doctrine on heterosexuality, against sexual pleasure, or against female sexual pleasure in particular. To do so in the name of inclusivity doesn't change what it undoubtedly is—a form of sexual violence. As J. K. Rowling has noted, it's harder to make the case that a trans woman is a woman when she acts like a man plagued by (actual) toxic masculinity.

The political and cultural commentator Andrew Sullivan—who is gay—speaks about all of this in terms that are applicable to each of us if we just substitute our own sexual orientation for his. Writing in 2022, he explained how disconnecting sex from sexual orientation is, in effect, existential for people who are gay. If you're heterosexual, bisexual, or lesbian, substitute "we" and words that describe your different sexual orientation for Sullivan's "gay male," "man," "male," "maleness," "gayness," and "homosexuality." It will read the same for you:

> One of the core elements of gay male culture—the celebration of the male *body*, its unique qualities, and its sexual power—is effectively diminished. It's diminished because we are told that being a man is now a feeling inside your head rather than a fact about your body. . . . All that gay male physical sensuality—the interaction of male bodies with one another, the passion for biological maleness— is reduced to an arid, gnostic, inside "feeling" unrelated to the body at all.

He concluded, again in broadly applicable terms if we substitute our sexual preference for his:

> At some point, gay men need to face down those who deny the biological differences in the human body that make homosexuality possible. If there is no sex binary, there is no homosexuality. We are not some third sex; we are one of two sexes: men. Our sex is not just in our head; it was not merely assigned at birth. It is in our bodies and minds shaped by testosterone since the womb, bodies that seek sex and intimacy with other male bodies shaped by testosterone in the womb. . . . That some gay rights leaders are now telling gay men they should force themselves to be attracted to vaginas . . . is an outrage.

I have a feeling that some in the forefront of this revolution know there is a large body of silent opinion among gay men and many lesbians that deeply believes in sex differences, cherishes and celebrates the male and female bodies, and does not see gayness as connected to transgender experience (which is not to say that transgender experience is any less valid). Gay happiness depends on our *owning* our own sex, not denying it. And biology is our friend.

Assuming real, not politically coerced, consent, sexual pleasure as Sullivan describes it is good. It's good beyond its connection to procreation because basic physical intimacy promotes emotional intimacy, which contributes to well-being and happiness. Given that well-being and happiness affect our health, it shouldn't come as a surprise that researchers find that sexual pleasure has medicinal effects. It can, among other things, reduce pain, relieve stress, improve sleep, lower blood pressure, strengthen heart health, and ameliorate menstrual cramps. More broadly, sexual pleasure is a natural part of a flourishing and fulfilled life. In Hinduism, sexual pleasure is "one of the highest and purest pleasures that God had given to the human." The millennia-old *Kama Sutra*, literally a text on sexual pleasure, was written "as a guide to the art of living well, the nature of love, finding a life partner, maintaining one's love life, and other aspects pertaining to pleasure-oriented faculties of human life." Dissenting in the 1986 Supreme Court decision in *Bowers v. Hardwick*, which upheld a Georgia law criminalizing sodomy, then seventy-eight-year-old Justice Harry Blackmun agreed, noting that "only the most willful blindness could obscure the fact that sexual intimacy is a sensitive, key relationship of human existence, central to family life, community welfare and the development of human personality."

I don't want to belabor Blackmun's community welfare point, but it's important to flag it—especially since sexual intimacy is first and foremost private business and so we either resist the suggestion that the community is somehow invested, or we just don't think about it. Part of this is that healthy individuals contribute to healthy communities. The other part is political. Governments that have some version of "life, liberty, and the pursuit of happiness" as an aspect of their corporate identity generally include sexual intimacy in that set of values. "It is not simply a matter of personal privacy versus the public interest" or leaving people be, wrote the Supreme Court of India in a case in which it decriminalized sex among consenting adults whatever their sexual orientation: "The modern perception is that there is a public interest in respecting personal privacy." Our intimate choices and decisions "must be secured against undue intrusion by the State because of the role of such relationships in safeguarding the individual freedom that is central to our constitutional scheme."

Sex Is Beautiful and Beauty Is Good

I've discussed sex for procreation and sex for pleasure. But sex is also beautiful, and beauty is itself good. Defining *beauty*, like defining *sex*, is as easy or difficult as you want to make it. I don't want to make it difficult and so here are the basics, from the New World Encyclopedia:

> Beauty is commonly defined as a characteristic present in objects, such as nature, art work, and a human person, that provides a perceptual experience of pleasure, joy, and satisfaction to the observer, through sensory manifestations such as shape, color, and personality. Beauty thus manifested usually conveys some level of harmony amongst components of an object. . . . Beauty has been recognized as a core value throughout history and in diverse cultural traditions. While beauty has cross-historical and cross-cultural recognition, the senses and the standards of beauty differ from one period to another, as well as from one cultural tradition to another.

The key points:

- Beauty is an important and universal experience.

- The experience of it is perceptual and sensory—pleasure, joy, and satisfaction.

- It involves harmony, which is often but doesn't have to be mathematically perfect.

- It's both individually experienced and culturally relative. In other words, it's in the eyes of the beholder and those eyes are influenced by nature and nurture.

Notions of physical beauty have consistently included both men and women. We sometimes find value in hierarchies of beauty, as in the most beautiful man or the most beautiful woman, but one of the magical things about beauty is that it isn't a zero-sum game. We can and often do know it in the moment without a comparator.

Of course, the sexed body is just *one* of the things we find beautiful about humans: A person can be beautiful for reasons having nothing to do with their physical form as male or female. A body can be beautiful that's covered—by clothes, tattoos, cosmetics—so that what we see isn't so much the physical form as personality. A clothed or unclothed body can be beautiful that's blended the physical binary. Again, it's not a zero-sum game.

Including the sexed body—the male and female form—among the things we experience as beautiful is naturally adaptive as it's related to sexual pleasure and procreation, but it's more than this. As with beauty generally, the experience is perceptual and sensory. We see and react with pleasure, joy, and satisfaction to the harmony among its parts, to its inherent integrity. The experience can also be intellectual, moral, or religious, as we appreciate the body's life-giving power and those who work to perfect their natural talents and gifts. Finally, it's plain from what we save, display, and celebrate that beautiful men and women are valuable beyond our individual sense of this. The best examples come from art and athletics.

The renowned art historian Kenneth Clark described the human body as the most "immediately interesting object" so that even as our "sense of the 'ideal nude' changes"—as to its form and its state of undress—the nude itself is "enduring." I began this chapter with a discussion of Michelangelo's *David*, carved in the early 1500s but inspired by sculptures from Greek and Roman antiquity. It's a perfect example of our constant fascination with the male form. The *David*'s female counterpart is arguably the *Venus de Milo*—though her lower half has always been adorned by marble drapery and her limbs have been lost to time. Discovered in 1820 on the Aegean island of Milos and probably sculpted by Alexandros of Antioch between 150 and 125 BCE, she is today in the Louvre and arguably more famous than the *David*. The two are often paired because both are seen as doubly beautiful, first as art and second as man and woman.

Like the *David*, the *Venus de Milo* is fascinating to viewers. That the fascination was political for France is the stuff of modern legend, as her acquisition and display in Paris were seen by the government as symbolizing the country's connection to empire and its devotion to timeless beauty. For tourists, she's been a constant destination, always a highlight of a visit to the City of Light.

As artistic inspiration, Venus is nearly unparalleled. The great modern sculptor Auguste Rodin wrote an essay, complete with sketches, about the *Venus*'s perfection. Contra Vasari on the *David*, Rodin proclaimed that "[n]othing can ever take the place of *this* persevering study" (my emphasis). He rhapsodized about her "sincerity to nature" and "the long curving lines of the back and hips" that "melt into each other in melodious harmony" before concluding that "approaching her step by step one persuades oneself that she has been modeled by the continuous washing of the sea." Salvador Dalí made a *Venus de Milo with Drawers*—a copy of the sculpture with, yes, drawers that open thanks to furry pulls. Miles Davis rendered her in jazz on his 1957 album *The Birth of the Cool*. In 1990, Jim Dine set up a bronze trio of Venuses *Looking Toward the Avenue* in midtown Manhattan.

This list could go on and on. It's perhaps best to end with Greg Lansky's provocative 2023 *Algorithmic Beauty*, a marble riff on the Venus as Kim Kardashian. Taylor Hunt described Lansky's work in an article called "Reimagining Venus de Milo in the Modern Age": she "stands at the same height as its antecedent" and "proudly displays plastic surgery scars while extending an arm with an iPhone in a clutch." If there was ever proof of Rodin's insistence that the original *Venus de Milos* is true to nature, not idealized, or of the broader principle that the ideal female nude is culturally relative, *Algorithmic Beauty* is it.

I've focused on the *David* and the *Venus de Milo* because they show that, even as our sense of the ideal form has changed over time, we've constantly found beauty—a perceptual and sensory experience of pleasure, joy, and satisfaction—in the harmonious male and female forms. But, of course, there are others. Among these are works that celebrate the beauty of the athletic body.

Beginning with the *Discobolus* (perhaps the most famous of all sports sculptures), through the fourth-century mosaics of female athletes in "bikinis" (discovered at the Villa Romana del Casale in Sicily), and the *David* (described by Paoletti as representing "the athletics of war"), to modern sculptor Robert Graham's dual figure sculpture *Olympic Gateway* (featuring a male and female nude in the aftermath of competition), art and the athletic body have constantly intersected.

We celebrate athletes for their effort and champions for their victories, and we appreciate the competition that it took to establish the hierarchy, but this is often inextricable from our celebration of the body—including the work on the body it took to get there. Most elite athletes embrace the fact that their bodies are their business. Our hearts and lungs are engines, and we train them as such. Our bones, muscles, tendons, and ligaments are parts that need to be strengthened and lengthened and coordinated, always symmetrically, or else the whole doesn't perform. Our adrenaline has to fire at the right times and, for females, our menstrual cycles may need to be controlled so as not to interfere with championship moments.

There's a stunning binary exhibit by the photographer Howard Schatz inside the Olympic Museum in Lausanne, Switzerland, that makes clear that each sport and event features a different set of physical excellences and that there is a difference between the male and female versions of each "perfect" body type: the basketball player and marathon runner; the gymnast and wrestler; the weight lifter and diver. Whether they're male or female, what these body types have in common is that they're sculpted by exceptional mental discipline and physical effort ultimately reflecting a body and mind in balance. The champion swimmer Cate Campbell put the point to me this way: "The elite athlete's body is beautiful because it's capable, it can do what it's been developed to do." Then, mirroring the Schatz photographs with her words, she added, "One of the best things about the Olympic Village is the dining hall where you see a parade of humans, everyone in the best shape of their lives, and their body types are so diverse. We'd play Guess the Sport. They're all beautiful."

Outside of the Olympic Village and the Olympic Museum, athletes, their promoters, and their countries take advantage of our tendency to experience their traits and athletic capability as physical beauty. On the commercial side among many others, see Cristiano Ronaldo. Gabby Reece. Coco Gauff. Cho Gue-sung. Politically, the three global superpowers are all in.

No country has taken this tool to the same heights as Russia, which has long specialized in cultivating athletic beauty in the strong, graceful, victorious version of that type: *this* is who *we* are. The ballet dancers. The gymnasts. The skaters. Kamila Valieva before the fall. As diplomats around the European Union wrangle over whether Russia should be allowed to take part in the 2024 Summer Olympics in Paris, the two best female high jumpers on the planet, reigning Olympic champion Mariya

Lasitskene of Russia and her heir apparent, Yaroslava Mahuchikh of Ukraine, are used by their respective countries as proxy instruments of war.

China borrowed from Russia's playbook around the 2020 Winter Olympics in Beijing, embracing freestyle skier Eileen Gu as "a symbol of a confident China." Gu, a dual national who grew up in California, used her social media posts to promote "her three lives as an elite skier, model, and top student who scored 1580 out of 1600 in her SAT." As Vice's Viola Zhou explained, "In a country where rigid gender roles put women in boxes and the government suppresses feminist activism, Gu is a rare example of a high-flying woman who has become a source of national pride but is also vocally encouraging young girls to break their limits." Since then, "the poster child for the 2020 Beijing Winter Olympics" has enrolled at Stanford and become an ambassador for the U.S. bids for the 2030 and 2034 Games.

The U.S. government is in on the action too. Its Sports Diplomacy division "is the best-kept secret in the State Department," according to its director. Star athletes are useful in spreading American values abroad, including the commitment to women's equality. In doing its work, the Sports Diplomacy office "has engaged with virtually every sport, including skateboarding, breakdancing, mountain climbing and baseball, although basketball and soccer are the most-requested ones." It shouldn't be a surprise that soccer star Megan Rapinoe has been the department's most requested athlete. She's politically controversial but also attractive and charismatic and because of this, enormously marketable. She's been on the cover of magazines from *Sports Illustrated* and *ESPN The Magazine*'s Body Issue to *Newsweek*, *Bazaar*, *GQ*, and *Glamour*.

Our transcendent fascination with the male and female form in art, in sport, and in their overlap is always subject to criticism. To the extent our fascination is natural, critics want to civilize it. To the extent it's culturally constructed, they want to tear down the house.

For many conservatives, nudity is associated solely with sexual relations, and they focus on modesty and putting clothes—or at least fig leaves—on the body in public. The *David* has been a repeat target over the centuries. There were his original copper fig leaves in the sixteenth century and the plaster leaf made especially for Queen Victoria in the nineteenth. Today we have the age-based unfurling contemplated by a school in Florida that specializes in, of all things, classical studies: "We're not going to show the full statue of David to kindergartners," Barney Bishop III, the head of the school board, told Slate's Dan Kois:

> We're not going to show him to second graders. Showing the entire statue of David is appropriate at some age. We're going to figure out when that is. . . . Maybe to kindergartners we only show the head. You can appreciate that. You can show the hands, the arms, the muscles, the beautiful work Michelangelo did in marble, without showing the whole thing.

The *Art Newspaper* reports that in response, Cecilie Hollberg, the director of the Galleria, invited everyone associated with the school—including Bishop, the outraged parents who had started the fuss by complaining that their sixth graders were shown the statute without advance warning, the ousted principal who allowed that lesson, and the entire student body—to come to Florence to view the "purity" of the *David* in situ: "To think that David could be pornographic means truly not understanding the contents of the Bible, not understanding culture and not understanding renaissance art."

Progressives at times also focus on modesty as a way to protect women and girls from being objectified and sexualized. Their main interest these days, though, seems to be in taking down hierarchies—including hierarchies of beauty. They're especially provoked by beauty in perfected or idealized forms, which is criticized variously as body shaming, Eurocentrism, white supremacy, and, most recently, transphobia.

There is some merit to these critiques, especially as they concern the blurry lines between appreciation, sexualization, and exploitation. Women have benefited tremendously in terms of health, confidence, and societal success from more liberal mores, from being nearly as much at liberty with our bodies as men are. But we remain much more likely than males to be mistreated when we do. The solution, though, isn't telling women and girls they can't celebrate their bodies publicly for fear of getting sexually assaulted, or rejecting hierarchies of beauty as immoral, or engaging in reverse body-shaming of beautiful men and women—any more than it is a compulsory hijab. None of these solutions is necessary to meet the underlying concerns. Indeed, they serve only to steal beauty and subtract value from the world.

At the beginning of this chapter, I quoted from a book on Michelangelo's *David* by John Paoletti. It includes a chapter fabulously titled "Naked Men in Piazza," about the sculpture's political significance. The *David* was originally located in the Piazza della Signoria, right in front of Florence's equivalent of city hall. Like the statues of Confederate soldiers or a blindfolded Lady Justice standing guard in front of U.S. courthouses, a resplendent naked man stood guard in front of Florence's seat of power. Statues in government places always mean something to the community and to the individuals who would stop to see them there.

Paoletti explains that there were two kinds of naked men in the piazza in Michelangelo's day. There were those who violated "the civil order of the dominating religion within the state" and whose "naked body parts" were "tacked to gates and buildings" as a warning to lawbreakers. Then there was *David*, the warrior, not the king, "whole" in his "idealized nakedness" and "symbol of the intact and healthy Republican

112

state." As sculpture, the biblical hero also represented inner and outer beauty, and again, "the athletics of war"—all attributes the Florentine Republic was anxious to project: *this*, in all of its dimensions, is who *we* are.

That the *David* has always been a public sculpture with religious and political significance didn't mean that it wasn't also always an extraordinary personal experience for all of the reasons you know if you've seen him for yourself. Our appreciation for the man in the marble can be purely intellectual, as for the representation of the biblical figure who was a warrior, a father, and a king, or as appreciation for those who cultivate or civilize their natural talents and gifts. But it's not always platonic. The carved stone exudes human desire. The biblical David loved Jonathan, and Michelangelo is also said to have been a lover of men; it shows in his work. The biblical David was also a lover of women—Bathsheba, with whom he had children, is only the most famous among them. Ultimately, the *David* captures all of this. It speaks to all of us because it "embodies human potential in its most vital form."

That embodied potential brings me back to Andrew Sullivan. We live in a moment in which, in some parts of the world and in some demographics, it's uncouth to speak the truths about sex. At least some of what I've written here is heresy in those circles, where we're told that we shouldn't be seeing them as truths and certainly not as good ones. Academia, my professional home, is sometimes a source and bastion of this alternate reality. But as Sullivan put it in the context of the assault on sexual pleasure, it's important to "face down those who deny . . . biological differences" and to continue to insist out loud that they are real and worthy of being celebrated.

5

Sex Just Is (Like Age)

THE FIRST TIME I MET Cate Campbell was in Budapest in 2022. The Australian swimmer is a four-time Olympian and multiple Olympic and world champion. Her specialty is the 50- and 100-meter freestyle, which means that on her best days, she's the fastest female on the planet, in the water. Campbell's tenacity and dedication, her brutal training regimen, and her consistency at elite level for over a decade have made her an unusually durable star. She swam in Beijing in 2008. In London in 2012. In Rio in 2016. In Tokyo in 2021. As I write, she's training for Paris in 2024. It's an understatement to say that Campbell is swimming royalty: in her world she's known simply as C1.

In 2022, we were both in Budapest to speak at the aptly named Extraordinary Congress where FINA (Fédération Internationale de Natation, now called World Aquatics), the international federation for swimming, diving, and water polo, was proposing a new eligibility rule for female competition and female world records. I had helped to draft the rule and was there to speak about our process and the federation's goals. Campbell was there to lend her support to the proposal.

The standards that were approved that day reaffirm the original understanding that the female category is about the physical body and whether it feminizes or masculinizes in puberty. They're sex-based at the outset—that is, eligibility is dependent on the athlete's chromosomal complement and the physical development that follows from that genetic foundation. They're then tied to the reason we have separate-sex sport (the insurmountable performance gap between males and females) and to the reason we specially protect the female category (elite sport's commitment to sex equality).

Whatever you may have read to the contrary about the performance gap, results data show that it's something we can take for granted. In Campbell's events, for example, there's not a single female among the top two hundred performers in the world, which is the cutoff for the online rankings. You'd have to go much further down the men's ranks before you'd land on the time that matches up to the female world record. Given this, fulfilling the commitment to sex equality in the elite sports space requires both separate-sex competition and eligibility rules for the female category that at least exclude competitors who have the male sex traits that drive the performance gap.

While there's debate in other sports settings about what sex equality means and requires, in elite sport we're not talking about giving women and girls the chance to try out for the team and to go home with a participation ribbon after we celebrate the victorious boys and men. Nor is it about access to the friendships and social skills development that youth and amateur sport often facilitate. Rather, in this space equality means one-for-one parity, including the same training, competition, and travel conditions; equal pay; and an equal number of spots in finals, on podiums, and in championship positions. Without this, at the Olympic Games in the 100 meters we couldn't know and publicly celebrate *both* a Michael Phelps (whose personal best is 47.77 seconds) *and* a Cate Campbell (52.03 seconds).

Going to Budapest, I knew about Campbell's athletic résumé but not about her personal journey or her related work on girls' and women's physical and mental health. She was born in Malawi to South African parents. When she was nine, her family moved to Brisbane, Australia, and it was the welcome of the local swimming community that gave her a home in a new and very different world. Nonetheless, she struggled with body image and with the management of the physical and physiological aspects of what she calls her "transition from girl to adolescent to woman"—even more so in her case for being in the glare of the public eye.

Like all females, Campbell says she had to "adjust to physical changes, fluctuating hormones, periods, changing body shape, all in a gendered context that tells girls that the way your body naturally develops isn't attractive and desirable if it's not skinny." Unlike most, she had to process all of this alongside elite sports practices—like skin-fold tests to measure fat—that are mostly based on a male model.

The male model assumes healthy male development "from boy to adolescent to man," including, from the onset of puberty, the rapid loss of body fat, an energy metabolism that favors muscle building over fat storage, increased joint stiffness, including in the knees and ankles, and increased strength, power, and coordination. As a result, adolescent males have "a consistent upward trajectory," Campbell notes. "They're always getting stronger and stronger in a linear way."

Healthy female growth is quite different, in some ways the opposite, geared as it is to fat development, an energy metabolism that favors fat storage, the establishment of the menstrual cycle, and the known direct and lesser-known indirect effects of that cycle. Among the latter is an increase in circulating levels of the ovarian hormone relaxin in the monthly pre-ovulatory phase, which contributes to ligament laxity (looseness) and changes the properties of tendons and cartilage. Because of all of this, Campbell says, when

> elite female athletes go from girl to adolescent to woman, our development isn't a linear upward trajectory. In the beginning there's an upward trajectory, but then there are the hormones and our cycles and our shape changes (breasts and bums and hips), and during that period our performance plateaus. When we're done growing and changing, *if* our body type is right and *if* we've been healthy about the process, then the upward trajectory resumes again and we become stronger.

Because male and female athletes train together their numbers and trajectories tend to be compared by coaches and sports officials. Even if the adults are careful, Campbell explains, "You inevitably internalize that boys' numbers are always better because lower means less fat, and less fat means more ripped, and faster times." She was talking about skin-fold tests but it could just as easily have been about the weight room. As a result, too many female athletes end up making unhealthy choices (or have them made for them) about training and food, which have negative implications for injuries and for energy storage, the latter with cascading effects on the menstrual cycle, period management, bone density, and so on.

Reflecting on the experience at a decade's remove from her own transition from girl to woman, she's clearly still stunned by the lack of attention to these fundamental aspects of female physical development:

> Like other elite athletes my body is my living, which involves 10ths and 100ths of a second differences between being on and off the team, on and off the podium, at or not at the top of the world. Because of this everything is meticulously planned and monitored. The

phases in the training cycle, the daily reps, my weight, my measurements, my sleep. But not our menstrual cycles, our development, our natural trajectories, and how they're managed? That makes no sense. We need to develop smart approaches to training females. The emphasis has to be on female biology.

Campbell is absolutely beautiful, inside and out. A statuesque, especially long-legged 6'1", at thirty-two years old she exemplifies the healthy version of the athletic aesthetic, neither the "super lean and ripped" version that prevails in sport nor the "fitness model" look that's the societal ideal. She told me, "I was never going to be super toned." Her gray-blue eyes and infectious smile top a strong, square jaw. Behind the face is a quick, secure, and thoughtful mind. She's been described as a "new breed of role model" and takes that role seriously as she navigates the intersection of sport and society:

Elsewhere, outside of sport, you're told to stay quiet, to stay small, to stay in your lane, all of which suggests that you're less strong, less capable. Physically less capable somehow translates to mentally less capable. People must think there's some biological connection between the body and the brain that justifies the patriarchy. It's important for women to see other women strong, loud, assertive, all different body shapes, and to see that celebrated as equal to men.

The privileges that I have now, I'm super aware that they're because women before me have fought for them. I have a responsibility to continue their work. It's a cliché but it's true that you can't be what you can't see. You can't know what you can do with your body before you see it done by someone with a body like yours. Representing the ultimate in what the female body is capable of through my performances and my presence on the podium is incredibly important.

When Campbell spoke to the delegates in Budapest in June 2022, all of her expressive power was on display. She began by naming the cornerstones of sport: inclusion and fairness. Recognizing that sport, broadly speaking, comprises "community or amateur sport" and "elite sport," she said that inclusion is a particular focus of "community or amateur sport," which "draw[s] people together, irrespective of background, race or religion, while also fighting the growing obesity and mental health epidemics." "Elite sport," though, is exclusionary by definition—a process of getting to the single person or team as champion is its bread and butter. Campbell explained that its existence is inextricable from "fairness" so that this has to be the focus of those who govern this distinct territory.

> Creating a place where men and women can come up against the best of their contemporaries and battle it out—down to 100ths and 10ths of a second. This battle, this standing up and comparing of wills and physique is what draws people to watch sport—to see who can squeeze the very last ounce out of their bodies and minds and emerge victorious. Without fair competition, sport in its elite sense would cease to exist.

She acknowledged that she has personally benefited from "fair, elite competition"—"I stand before you, as a four-time Olympian, a world champion and a world record holder"—but also signaled through her words a shared understanding that her accomplishments and stature have broader value. Hers was not merely a self-interested position.

Embracing the version of fairness that includes sex equality as the one-for-one proposition I described above, the one that allows Campbell and other female athletes a podium of their own to stand on, she rejected as too facile the argument that in elite sport we can achieve "balance" between the cornerstone values of fairness and inclusion.

That men and women are physiologically different cannot be disputed. We are only now beginning to explore and understand the origins of these physiological differences and the lasting effects of exposure to differing hormones. Women, who have fought long and hard to be included and seen as equals in sport, can only do so because of the gender category distinction. To remove that distinction would be to the detriment of female athletes everywhere.

Speaking directly to gender-diverse athletes at the conclusion of her remarks, Campbell said, "We see you, value you, and accept you." There was no denial of their existence or their right to take part in all aspects of the sport. She acknowledged that the form that this acceptance was taking would be hard on those who had hoped for more. Being seen, valued, and welcomed into the elite ranks like everyone else, in the category that corresponds to their sex, wasn't what certain transgender women and their allies were looking for. She called for "all the federations sitting within this room to examine your own policies to ensure the world of swimming remains inclusive." I later asked her what she meant by this, and she responded:

> We should acknowledge and celebrate both females and transgender women. In our push for equality we've forgotten about diversity. It's not a competition, it shouldn't be a competition, because we don't have the same physical experience. If we discount sex we discredit the experiences of 50 percent of the population. We discredit the female experience. We need to distinguish between sex and gender: sex is by its nature binary, gender is not.

Back in Budapest, she ended her speech by coming full circle to her own journey from community and amateur to elite sport:

It is my hope that young girls all around the world can continue to dream of becoming Olympic and world champions in a female category prioritizing the competitive cornerstone of fairness. However, it is also my hope that a young gender-diverse child can walk into a swimming club and feel the same level of acceptance that a nine-year-old immigrant kid from Africa did all those years ago.

The New Science of Sex Differences

Beyond the junk science that characterized many of the dumb pronouncements about sex in the late nineteenth and early twentieth centuries, two key discoveries were made that revolutionized our understanding of the biology of sex. First, scientists discovered and characterized the functions of our sex chromosomes, X and Y. Then our main sex hormones—testosterone, estrogen, and progesterone—were discovered and described. Together these discoveries laid the foundation for future scientific and medical research on sexual differentiation and development—including on DSD and sex differences more generally. It would take a while, but they also laid the foundation for all of us—not just the research community—to become sex-smart.

Before the late twentieth century, there wasn't a great deal of interest in—and so no real funding for—research into how males and females are different beyond the fundamentals of sexual development. Work on human anatomy, physiology, function, and disease continued, of course, but on the male model. This almost singular focus and the corresponding erasure of the female model was rationalized on the historically ironic grounds that we're alike enough that it's okay only to use males as research subjects—and to assume that the typical human is the average male. The fact that the menstrual cycle makes it more difficult to study the female body—and that it's more expensive to study both and then contrast the two—helped to lock in this focus even as it was obviously inconsistent with the premise of sameness to take this position. As a result, researchers did such famously inane things over the years as to study breast cancer—which primarily affects females and is tied to estrogen and progesterone production—using only male subjects.

The seeds of change were planted in the early 1990s when the NIH began requiring that both sexes participate in human research. But this initial effort fell short because the NIH didn't require researchers to compare males and females, or to analyze enough participants of each sex to be able to establish whether there were differences in the ways male and female patients with the same condition present, or the effects of sex on the safety and efficacy of a drug or treatment regimen. It wasn't until 2014 that the NIH required that all animal research consider sex as a biological variable. This led to an explosion in work directly comparing the two sexes to establish whether significant differences exist.

A major milestone along the way between the two NIH decisions was the publication of the report *Exploring the Biological Contributions to Human Health: Does Sex Matter?* (2001). The report was the product of a study group appointed by the Institute of Medicine (IOM)—now called the National Academy of Medicine—based on a comprehensive review of the best available evidence to that point. The IOM's "overarching conclusion" was simple: "Sex matters." From there, its key findings and recommendations were as follows:

- "Every cell has a sex" and so research on sex at the cellular level should be promoted.

- Sex begins in the womb, and so sex differences should be studied "from womb to tomb" when necessary by "min[ing] cross-species information."

- Sex "affects behavior and perception" and so researchers should both "investigate natural variations" and "expand work on sex differences in brain organization and function."

- Sex affects health, and so both "sex differences and similarities" should be monitored "for all human diseases that affect both sexes."

Consistent with the definitions we saw in chapter 1, the report defined *sex* as "the classification of living things, generally as male or female according to their reproductive organs and functions assigned by the chromosomal complement"; and it defined *gender* as "a person's self-representation as male or female, or how that person is responded to by social institutions on the basis of the individual's gender presentation." To ensure the integrity of sex differences research, it strongly recommended that scientists clarify their use of the two terms.

It also recommended supporting and conducting additional work on sex differences; making sex-specific data readily available; determining and disclosing the sex of origin of biological research materials; constructing longitudinal studies so that their results can be analyzed by sex; identifying the endocrine status of research subjects; and encouraging and supporting interdisciplinary research on sex differences to reduce the potential for discrimination based on identified differences.

The National Academy of Medicine operates pursuant to a nineteenth-century congressional charter as an "independent, evidence-based scientific advisor" to policymakers, including to the federal government. It took thirteen years for the NIH to adopt the recommendations in the IOM's *Does Sex Matter?* report, but when it did, it was a big deal because federal funding sets the agenda for research on humans around the world. Twenty years on, every field of medicine has sex-differences researchers and research being done that, per the NIH requirement, distinguishes on the basis of sex. Other specialists are able to mine the vast amounts of data developed on the federal dime to power their work.

Like almost all scientific and medical research, this work is focused on average, group-based differences—how we usually are. Some progressives criticize this basic scientific method as exclusionary (of people who fall outside of the norms) and thus morally wrong. For such critics, as we've seen throughout the book, anything that excludes—basically any drawing of lines among people—is immoral, and the very idea of atypicality is pejorative rather than simply statistical. As they see it, we should be focusing on individuals, especially those who are least well-off, not on the group. As applied in this context, the priority is on people who are sex or gender diverse, not on people who are some kind of normative.

These critiques, like most in the culture wars around sex and gender, are ultimately political. As I write, certain advocacy groups are arguing in federal court that sex-based policies like girls' and women's sport can't be based in "stereotypes"—by which they mean generalizations—about the strength, speed, power, and endurance of males relative to females. As applied to sex itself, this stereotype argument—specifically the suggestion that all sex differences are impermissible generalizations—has thus far been rejected by the Supreme Court. This is because sex matters in law as in life, and accurate generalizations are policymakers' bread and butter. How people usually are and what they usually do aren't the only factors that inform the development of rules and laws, but they are the biggest ones. Norms are always relevant, even if it's only to know how much it will cost to ignore or change them.

It's important that the scientists who are today working to improve our knowledge of average, group-based sex differences would be unrecognizable to the ignorant misogynists we met in chapter 2. Historically, too many scientists made nonsense pronouncements about sex differences not only because they lacked the sophisticated tools necessary to investigate important biological functions, but also because they were driven by patriarchal imperatives in the selection of their research questions, in their methodology, and in the results they sought to obtain. But with an entirely different mindset and much better research tools, today they usually do it really well.

My favorite example is a 2022 paper by neuroscientists Alex DeCasien, Elisa Guma, Siyuan Liu, and Armin Raznahan in the journal *Biology of Sex Differences* called "Sex Differences in the Human Brain: A Roadmap for More Careful Analysis and Interpretation of a Biological Reality." Borrowing from the *antiracism* movement—which asserts that to dismantle structural racism and deal with everyday racist occurrences, it's necessary to see race, to consider how it's being used, and to act affirmatively against racism wherever it appears—DeCasien and her colleagues argue for an *antisexist* approach to neuroscience research. They say that sexism exists not only when one sex or the other is seen as inferior in some way but also when we pretend—or manipulate research questions in an attempt to arrive at results suggesting—that sex doesn't exist. Being antisexist requires (1) rejecting *both* kinds of research projects and lines of inquiry; (2) pursuing research apolitically, using gold-standard methodology; and (3) limiting the misuse of good work by bad or careless actors and being clear and transparent about what the evidence does and doesn't show. They conclude that, "rather than avoiding, dismissing, or over-interpreting findings of brain sex differences, more accurate description could reduce the misrepresentation and misuse of such research, both within the scientific community and throughout society as a whole."

It's undeniable that sex-differences science has made huge leaps in the twenty-first century. We get a taste of the innovative research if we look at three additional areas beyond the systems that support human performance that I discuss throughout the book: the brain, the cardiovascular system, and the immune system. What you'll see is that each has sex differences with important functional effects tied, directly or indirectly, to age and reproduction. Although both the differences and their functional effects may be influenced by our environments, including our neonatal exposures, they all have a demonstrable biological basis in our genes and sex hormones.

The Brain

As we saw in chapter 1, although there's substantial overlap between the brains of human males and females, there are also average sex differences in brain size, composition, and physiology. Here are three examples, one in each category.

On average male brains are 11 percent larger than female brains, and after controlling for brain size, there are also average size differences in about 67 percent of their subparts. Using MRI technology, researchers have found that these differences are already present in utero and that at the population level, they can be used accurately to infer sex about 95 percent of the time. A focus of researchers has been on size differences in the male and female limbic systems—especially within the amygdala and hippocampus, which are involved with memory formation and recall and, relatedly, with our emotional and behavioral responses.

There are also average sex differences in the amount and location of gray matter (neurons) and white matter (nerves). White matter is white because of the myelin that coats the nerve fibers. Myelination occurs throughout early life, but at very different rates in males and females (more and sooner in females). Generally speaking, gray matter processes information and white matter communicates it among parts of the brain and between the brain and the body.

Finally, there are sex differences in brain metabolism—meaning the amount of circulating glucose and the way it's used. These differences exist in the brain overall and in its subparts. They are related to differences in metabolic brain age—by the end of adolescence, male brains are already metabolically older than female brains and that difference doesn't go away.

Consistent with the approach DeCasien and her colleagues recommend, it's important not only to know but also to say that these differences are small but nevertheless statistically significant—meaning they're real. People who tend to want to deemphasize or even to ignore brain sex differences use the facts that our brains are mostly the same and that a given individual's brain may not correspond to the average for their sex to suggest that the differences aren't worth noticing. This doesn't make sense. We know that tiny things can have important impacts—like a single microscopic sperm on a pregnancy—and that in the aggregate, even independently insignificant things can also matter, a lot. As applied here, we know that small sex differences in the brain operating as part of our neurological systems influence behavior, cognition, and disease patterns.

Behavioral examples include impulse control, which is related to myelination (it's estimated that it takes to age twenty-four before males reach the levels females have by age twelve); libido (which is markedly higher in males); and aggression (which is more likely to be intense and physical in males). Libido and aggression are both regulated by the hypothalamus.

Cognitively, there is almost complete overlap between males and females in terms of the ability to "reason, plan, solve problems, think abstractly, comprehend complex ideas, learn quickly, and learn from experience"—which is the way experts have defined general intelligence. We may use different pathways, but we usually get to the same place. On average, though, males are better at visuospatial tasks than females, meaning they are likely to be better at perceiving, analyzing, and manipulating relationships among objects. It's hypothesized that this advantage explains the male dominance in chess and makes sense of a protected female category here too. As the television show *The Queen's Gambit* illustrated, however, sexism is rife in this setting. Until that's sorted out, we won't know for sure.

In terms of disease patterns, the typical age of onset of many conditions also differs by sex, including autism spectrum disorders and attention-deficit and hyperactivity disorder (earlier onset in males), anxiety, depression, and migraine (adolescent onset in females), and neurogenerative disorders like Alzheimer's and Parkinson's disease (later onset in females).

The biological bases for these sex differences lie in our chromosomes and gonadal hormones. Sometimes working together, sometimes directly and independently, they contribute to the masculinization or feminization of the brain beginning already in utero. At that early stage, male brain development is strongly influenced by gonadal hormones whereas female brain development is mainly the product of genetic influences. Thereafter, as we saw with impulse control and metabolic age, the female brain develops not only differently but at a different pace than the male brain—mainly it matures two to several years earlier depending on the attribute. (This won't come as a surprise to anyone with sex-typical boys.) Although sensation-seeking behavior continues beyond this point, the female brain is basically mature by the beginning of the female body's prime fertility years—at around fifteen to sixteen.

As I noted in chapter 1, in the brief introduction to brain plasticity, it's understood that our experiences—including our intrauterine environments—are also influential. But as Art Arnold, one of the leading researchers in the field, put it to me, "it's often impossible to control variables in studies of humans" and so many of "the actual causal factors aren't established." This is in part why he and Margaret M. McCarthy, another leading neuroscientist, recently flagged exciting new developments in epigenetics and technology that are allowing for "the possibility of greater appreciation of the complex biological-environmental interactions that give rise to sex differences in [brain] physiology and disease."

In an exchange about sex differences in their field, three colleagues in the Duke Department of Neurosurgery, one of the top programs in the world, gave me this taste of the future. Quinn Ostrom, Kyle Walsh, and Mike Brown work on brain tumors, specifically gliomas—including glioblastomas and meningiomas. They explained:

Sex differences in both brain tumor incidence and prognosis have been reported, with females having a lower incidence of malignant glioma, as well as longer survival during standard of care therapy for grade IV glioma (i.e., "glioblastoma"). It remains to be determined whether differences in glioma incidence and outcomes are hormone related or not, as well as whether they are related to differences observed in the incidence or severity of other neurological diseases (e.g., multiple sclerosis and Alzheimer's).

Meningioma, a primarily benign tumor of the membrane that surrounds the brain, is diagnosed twice as frequently in females compared with males across the lifespan. Interestingly, the increased risk in females begins to emerge around the time of puberty and subsequently decreases after menopause, strongly implicating hormonal factors in meningioma initiation and progression.

It's all very exciting!

The Cardiovascular System

Virginia Miller is a physiologist who spent many years on the faculty at the Mayo Clinic. Now retired, she was one of the leading sex-differences researchers in the world among those focused on the cardiovascular system. Her work filled in gaps in our knowledge of the female heart, and that was invaluable given that heart disease is the number one killer globally of men and women and that there are sex differences in heart structure and function, stroke volumes, resting heart rates, and age-related changes in physiological variables, including blood pressure.

Blood pressure is generally lower in young women than in young men. Starting at about age thirty, it drifts upward in men throughout life. It changes more slowly in women until menopause, at which point there's a sharp increase in blood pressure. By age sixty or so, more women are hypertensive than men. The "hockey stick" pattern of blood pressure in women is due to the loss of female sex hormones at menopause. In premenopausal women, these hormones relax the blood vessels and inhibit the nerves, which cause the blood vessels to constrict.

As I'm sure you expect by now, sex differences in the cardiovascular system are understood to be biologically based and related to age and the reproductive imperative. As Miller explained in the *European Heart Journal* in 2020,

> The bodies of females and males differ at the most fundamental genetic level due to the presence of the sex chromosomes. The genes on these chromosomes not only direct development of the reproductive organs with subsequent production of sex steroid hormones, but also influence other organs and expression of genes on the autosomes.

In other words, as sex differences in human performance and the brain are influenced by genes and hormones acting separately and together, so are sex differences in the cardiovascular system. Genetic and hormonal differences drive the structural and functional differences that ultimately "allow for the female cardiovascular system to adapt to changes necessary to sustain a viable foetus, including increases in blood volume, autonomic regulation of blood pressure, and cardiac dynamics, i.e., general cardiovascular function." An individual female may never gestate a child, but her cardiovascular system is designed in part to support this reproductive process.

Finally, it's understood that biologically based sex differences in the cardiovascular system join with environmental factors over time to influence "sex differences in incidence, prevalence, morbidity, and mortality in cardiovascular disease between males and females."

Marking the twentieth anniversary of the publication of the Institute of Medicine's *Does Sex Matter?* report, Miller reminded the research and clinical communities that they have an ongoing "responsibility to account for sex as a biological variable and to understand that hormonal status changes throughout life"—*and* to take both "into account in the design of basic science studies into the mechanisms of disease as well as the development of novel therapies." Both old-school research designs that don't bother to disaggregate sex and gender and new-school designs that intentionally conflate the two, for example by coding as "male" not only males but also females with male gender identities, reduce the granularity that matters for everyone.

The Immune System

The immune system is a made up of a coordinated group of organs, cells, and substances whose role is to balance offense (mounting a successful attack on invading organisms) and defense (ensuring that the battle doesn't destroy its own cells and systems). Militaristic language abounds in immunology. The immune system works in part by remembering the germs it's previously encountered and successfully defeated, so that it can mount a more efficient and effective response if it encounters a familiar version in the future.

As always, defining *sex* as *biological sex*—here specifically "the differences between males and females caused by differential sex chromosome complement, reproductive tissues, and concentrations of sex steroids"—researchers have developed "robust" evidence of biologically based "sex-specific differences in overall immune function." Many of the differences are tied to age, especially to aging, which is "associated with reduced antiviral immunity and control of virus replication." On the disease side, differences range "from autoimmune diseases and allergy-induced asthma to microbial infection and cancers." On the treatment side, they include "response to vaccines, antibody therapies, and immunotherapy." In general, females "mount more effective cellular and humoral immune responses and are less likely to succumb to bacterial and viral infections than men."

The Covid pandemic brought the female bias in immune system function to the front pages of newspapers around the world. In 2022, the *New York Times* noted, "It's one of the most well-known takeaways of the pandemic: Men die of Covid-19 more often than women do." As microbiologist Sabra Klein put it, "You can't attribute observations about things like mortality from a complex disease like Covid and say it's all biology," but "I also don't think you can say it's all social and it's all behavioral, either." Klein, a leading figure in the study of sex differences in immunology, explained that modeling based on electronic health records showed the "disparity could be substantially accounted for by greater inflammatory responses among men."

The greater immune response among women is both "a blessing and a curse," Klein told *The Wall Street Journal* in 2021. It helps "explain why Covid-19 tends to be less deadly in women than in men" but also why women are more likely than men to suffer from long Covid and from certain vaccine side effects. The overall Covid pattern is consistent with "male-biased severe outcomes from viral infections," including for "seasonal influenza viruses, SARS and MERS viruses, hemorrhagic fever viruses, hepatitis B and C viruses, and herpes viruses." It is also consistent with female-biased severe outcomes from autoimmune diseases. There, a striking 80 percent of the patients are women.

• • •

It turns out that sex is also in our skin. As is the case throughout the body, this molecular difference has functional effects that are understood to be tied to reproduction. Universally, females after puberty "carry a lighter skin than males." In general, skin color is "an adaptive trait resulting from the trade-off between protection against ultraviolet radiation (provided by darker skin tones) and production of vitamin D (enhanced by lighter skin tones)." The prevailing explanation for the sex difference in "brightness" is that "darker skin pigmentation" in males "optimizes the folate levels in the body" and these "safeguard sperm production." In contrast, "females need to maximize the synthesis of vitamin D during pregnancy and lactation to increase their infant's and their own biological fitness."

Which brings me back to Cate Campbell. In between the Rio and Tokyo Games, she was diagnosed with melanoma in a mole she'd had from birth. Melanoma has been dubbed "Australia's national cancer" because it "kills one Australian every five hours." The sun blazes in the Land Down Under, where there are a lot of lighter-skinned people whose melanin levels aren't calibrated for the climate. But melanoma is the most virulent and deadly of skin cancers everywhere. This includes Malawi, where Campbell is originally from and which could benefit from research that didn't assume darker-skinned people are somehow immune.

Melanoma is personal for me because my very light-skinned mother, Bluette, died of the disease in early 2002. When she was first diagnosed, it had already crossed her skin barrier and invaded her body, and there was no cure then for metastatic melanoma. The only question was which of her organs the disease would capture first, because that would be the one to end her life. About a year later, we got the news that it had metastasized to her brain. She had grown up in Lausanne, where very few kids of her generation went to university and most were tracked early into trades. At fourteen, she was apprenticed to and ultimately became a dressmaker, one of the few creative possibilities available to Swiss girls at the time. It was what she did until she could no longer thread a needle. She never had any formal higher education, but throughout her life she was fascinated by science. I once asked her how she felt about having taken an apprenticeship so early. She said her trade had served her and our family well, but that in another life she'd have liked to have been a brain scientist. When she was offered the possibility of being a research participant in a melanoma study that was being conducted out of the Dana-Farber Cancer Institute in Boston, she jumped at the chance. She knew that her participation wouldn't be therapeutic for her and that she wouldn't live to know what the research would eventually teach us, but it gave her final months added meaning to know that even though she had been "only a dressmaker" she had contributed to brain science in this life. My mom was unbelievably strong and independent to the end. The final trip she took, on her own just three weeks before she died, was on a train from New York to Boston to visit her study's principal investigator.

Twenty years on, the work done out of Dana-Farber and other research programs around the world has allowed scientists to home in on the immune system as central to melanoma's disease pathway and on immunotherapy as treatment. Consistent with the fact that it's an immune-related disease, the work has also shown that "biological sex is a fundamental factor in melanoma." Getting the disease, how it develops, and whether you survive it are different on average depending on whether you're male or female. As you might predict, these sex differences interact with age. Before "50, females actually have approximately twice the incidence of melanoma as males, but thereafter this difference changes, and by the age of 60 males have twice the incidence of females and by age 70 this difference rises to three times as great for males." In "terms of sex differences in survival from cancer" generally, "cutaneous melanoma is one of the most striking" as "females retain a 38 percent survival advantage compared to males after adjusting for stage at diagnosis." Finally, also predictable is that scientists who study sex differences in the disease are looking at the range of biological to environmental correlations, including *both* chromosomal *and* gonadal hormone effects and "behavioral-related pathways such as differences in UV exposure, primary care access, and skin awareness." I can still hear my mom with her strong French accent insisting, even though we aren't light-skinned like her, "Sunscreen!"

Thankfully, Cate Campbell caught her melanoma early. Since her treatment, she's added skin cancer prevention and survival to the set of causes she champions from personal experience. She's also all about sunscreen.

6

Sex Is Still a Problem
(Like Race)

FOUR MONTHS IN 2022 made clear that even in the developed world, where people go to grad school, conceive start-ups, park their brains on TikTok, attend concerts, and think of themselves as members of a liberal global community, sex still matters in ways that are human-made and unnecessarily destructive.

On February 24, Russia invaded Ukraine, sending thousands of young men to the front lines. The drafts instituted by both countries in the wake of the invasion were a stark reminder that at the end of the day, individual males are presumed, simply by virtue of the fact they were born into a male body, to have consented to leave behind their lives and families and risk death in battle. In wartime, young males as a group have no higher secular purpose.

Exactly four months later, on June 24, the United States Supreme Court issued its decision in *Dobbs v. Jackson Women's Health Organization*, overturning *Roe v. Wade* (1973). In *Roe*, the Court said that at least in their first trimester of pregnancy, women and girls have the unfettered right to choose to continue or terminate a pregnancy. States exercising their police powers could give women more leeway but not less. *Dobbs* took that liberty back—*it was never there to begin with*, the Court explained—reminding us that at the end of the day, individual females are presumed, simply by virtue of the fact that they were born into a female body, to have consented to reproduce and rear children. We are communal property to those ends.

It's an understatement to say that *Roe* revolutionized women's possibilities. Together with the cases upholding the right to use contraception—*Griswold v. Connecticut* (1965) and *Eisenstadt v. Baird* (1972)—it meant that our lives were no longer determined by our procreative role, or by men who would deny us choices about the consequences of sexual relations and rape. We could now plan when, or whether, to have children, and the number that fit our circumstances. We could rely on abortion care as health care when continuing a pregnancy would cause physical or emotional harm. Or when it was ending naturally to ensure that medical complications wouldn't cause further damage. This last point isn't trivial: 10 to 25 percent of pregnancies end in miscarriage in the first trimester, more earlier than later, and pregnancy and childbirth are still fatal for many women. As I was working on this chapter, the beautiful sprinter and multiple Olympic gold medalist Tori Bowie died at home, alone, in her eighth month of pregnancy, delivering her daughter Ariana, who also didn't survive the event.

As *Roe* revolutionized the prospects for individuals it also contributed to the nation's prospects. As Treasury secretary Janet Yellen put it in 2022, in remarks that echoed Myra Bradwell's from the 1800s, societies are stronger—including economically—when the law empowers women and girls to be full and equally free citizens in the political community.

Nevertheless, when the Court released the *Dobbs* decision, states that had been chomping at the bit to restrict or abolish abortion acted quickly to exercise their new authority. Thirteen states had "trigger" laws in place that would make abortion automatically illegal in the event of a favorable high court ruling. Ohio's version banned abortion after six weeks, about the time a fetal heartbeat is detected. Several weeks before the June 24 decision, a ten-year-old girl living in Ohio was raped twice by a twenty-seven-year-old man. In late June, she was taken to Indiana, where it was still legal to have an abortion after six weeks. For those conservative lawmakers who believe that life begins at conception and that it's the bearer's obligation to protect it, the girl's age and circumstances—what it would do to what's left of her childhood, to her developing body and brain, and to her life going forward—were all beside the point. Because her rapist had deposited sperm in her vagina that joined with an egg in her uterus to create a human embryo, her life was to be determined by his crime.

These new laws are being challenged, mostly on state law grounds. As I write, Ohio's six-week ban has been enjoined by the lower courts as they decide whether it's permissible under the state's constitution. But the bottom line is that in many parts of the United States, women and girls are back in circumstances not known for fifty years: without the liberty in their first trimester to decide whether to bear a child. Of course, in other parts of the world, this right never existed.

Many of the images we saw and stories we read in the earliest months of Russia's war against Ukraine were of Ukrainian women fleeing the country with children in cars, on trains, on foot, to cross borders into neighboring states that would take them in as refugees. A caption accompanying a set of devastating photographs from the Polish border by NPR's Ben de la Cruz read "Most of the nearly 3 million Ukrainian refugees are women and children." One image is of three generations in what looks to be a single family, a young girl, an adolescent girl, a middle-aged woman, and an older woman. Since time immemorial, women have done this with their children when the battle draws near.

Most Ukrainian women remained behind, of course, and some enlisted voluntarily before the country's draft was formally extended to include females. But as is the case with women in the military in other countries, most serve in support roles, not on the front lines. I remember listening to the BBC on my drive to school one day during the first year of the war and hearing a trigger warning. The Brits aren't soft and so this got my attention. Amid the grisly details, I heard Valeria, an assistant anesthetist, saying, "I have the most amazing job in the world. I defend heroes." "They defend us and I'm here to defend them—and not to let them die." For her, "the worst part of the job is when a 'construction set' arrives, soldiers' body parts that must be matched and placed together for burial." But "the youngest casualties" are those she won't forget. "When there is a date of birth of 2003, you realize that this person is 18 years old. This person saw very little in life, maybe never kissed and already sees death, sees and endures such severe trials."

Like most other countries, the United States has only ever drafted males. Men's groups have challenged this as sex discrimination, so far unsuccessfully, and advocates for women's advancement in the military have moved the needle on the possibilities for voluntarily enlisted females serving in combat roles. But there remains deep reluctance on the part of policymakers fully to embrace either idea. Two of my female students, Chandler Cole and Johanna Crisman, graduated from the United States Military Academy at West Point and served in combat roles overseas. They coproduce the *Combat Exclusion* podcast, where they raise awareness about gender issues in the military. They explained to me that while some of this is just antiquated thinking about what females can do physically and men can accept culturally, it's also related to sexual violence. This was reflected in Ukraine after the invasion: some of the women who chose to stay behind, or had no choice in the matter, were raped by the invading army. This is nothing new. You've surely heard the line "to the victor and the captor go the spoils." What you may not have focused on is that "the spoils" include women and girls. It's why "rape as a crime of war" is a legal term of art. Putting females into combat roles where capture is a real risk makes the generals and politicians nervous.

The modern twist on this otherwise timeless story was Ukraine's surrogacy industry, which is said to be among the best in the world. As Russia began its invasion, would-be parents from abroad tried desperately to get newborns and pregnant surrogates out of the country. But as Isabel Coles of the *Wall Street Journal* reported in March 2022, often "[w]e do not know where these unfortunate women are."

I've argued in this book that sex sometimes matters because it's good: we need male and female bodies to have kids and sexual pleasure, and they add beauty to the world and all of this is valuable. I've also argued that sometimes sex matters not because it's good or bad but because it exists as an immutable physical condition, like age: biologically based sex differences in the human body beyond our obvious reproductive differences are real and they have functional effects that are important to our lives. But there are also times that sex, like race, matters because it's the basis for unnecessary harm.

Mostly today we're not talking about formal, de jure, discrimination like the laws of abortion and war. What's much more common are formal policies, informal norms, and social developments that are facially neutral—they say nothing about sex one way or the other—but they impact males and females differently. The number of things in our daily lives that track sex is limitless: Trends in poverty track sex. Trends in single parenthood track sex. Trends in kids' ability to rise out of poverty are linked to their mother's relationship status. Educational and work-force participation rates track sex. Avoidable mental and physical health conditions track sex. In general, as the authors of "Sex Differences in the Human Brain" noted, there is

> extensive evidence that individuals who are outwardly perceived as male or female experience pervasive differences in diverse domains of life, spanning the educational, professional, and medical arenas, in addition to risk of exposure to different dangerous situations.

Intent matters, of course. Sex discrimination that formally privileges one sex over the other, or that's either designed or nearly certain to cause sex-specific harm, has a moral edge to it that causes special pain. I'm thinking about women and girls in Afghanistan who can't go to school or to the park because they're physically consigned to the home and, again, about military drafts the world over that consign young men to war. But people have to live with harmful impacts even when they're the result of merely negligent choices: those that are made because of the things we don't bother to know or see. Caroline Criado Perez's essential book, *Invisible Women: Data Bias in a World Designed for Men* (2019), is chock-full of examples. At the end of the day, all of it cuts.

For all of recorded history, laws and norms have been harder on women than men. However we identify, we're treated and experience the world differently because of our reproductive biology and inferences about us because of that biology. Because we're half the population and fill important societal roles, many of these policies and norms cause further harm to our families, communities, and countries.

Because of this overarching disparity, I'll focus on women and girls and specifically on three broad phenomena that account for a lot of the unnecessary—that is, avoidable—harm they experience today: sexual and domestic violence, caregiving norms, and the mental health epidemics associated with female puberty. I'll close out with the most important avoidable harms experienced by boys and men.

Sexual and Domestic Violence

People of both sexes, whatever their sexual orientation and gender identity, are at risk of rape and other forms of sexual and domestic violence. But there's no question that females disproportionately suffer from this special category of harm.

Definitions must be specific or there's no way to make sense of the statistics and records. *Rape* is usually defined as forcible penetration of the vagina or anus by a penis, that is, as nonconsensual intercourse. Modern definitions sometimes include penetration of the mouth and add digits or objects beyond the penis, but the traditional sense of *rape* still contemplates that particular organ and sexual gratification of the male body. *Sexual assault* is a broader term that includes rape as well as any nonconsensual sexual act or touching, including kissing. *Domestic violence*, according to the United States Department of Justice, is "a pattern of abusive behavior in any relationship that is used by one partner to gain or maintain power and control over another intimate partner." People usually use the term to mean physical and sexual violence, but it can also include economic, psychological, and technological abuse.

The numbers vary depending on the context, but about 90 percent of victims of sex-based violence are female. The real proportion of male victims may be higher than 10 percent given the different stigma associated with reporting that implicates weakness or homosexuality. The numbers of male victims grow, too, wherever the broader definition of sexual assault—"any nonconsensual sexual act or sexual touching"—is used. Still, there's no question that sexual and domestic violence are gendered crimes.

Age and culture also play an important role. Females are more likely to be subject to rape and sexual assault when they're between the ages of sixteen and nineteen. When they have intellectual disabilities. When they live in Botswana, where an insane 96.87 percent of females—almost every single one—are said to experience rape at least once in their lifetimes. When they live on Indian reservations, which have the highest rates of sexual and domestic violence in the United States. When they live in parts of the world where there aren't safe sanitation facilities. In the slums in Mumbai, for example, there are about "six bathrooms for 8,000 women" and a corresponding "danger of sexual assault from men who lurk near and on the routes to areas which are known to be used by women when they need to relieve themselves." As UN secretary-general Ban Ki-moon noted in 2014, not about Mumbai in particular, "it is a moral imperative" that we "ensure women and girls are not at risk of assault and rape simply because they lack a sanitation facility." This isn't an exhaustive list, just a depressing one.

The perpetrators of sexual and domestic violence are almost always males who take advantage of their relative strength, speed, and power. Some perpetrators of sexual and domestic violence are females who prey on their partners, and some are males who prey on males, but these incidents are not the norm. Speaking to domestic violence specifically, the British organization Women's Aid explained that "90 percent" is "caused by men within heterosexual relationships."

I'm fortunate never to have been raped, but like many if not most women and girls, I have been sexually assaulted as that term is now broadly defined. As a five-year-old when a gaggle of slightly older boys decided they wanted (just) to see a vagina. As a seven-year-old when a different gaggle of older boys decided they wanted (just) to mimic humping behavior they'd seen somewhere in their lives. As a sixth grader when a man decided (just) to expose himself masturbating as a classmate and I were waiting for a city bus to take us to school—that made for my first experience in a courtroom. As a twelfth grader by the single father of a child for whom I was babysitting while he was on a business trip. Returning home in the middle of the night a couple of days earlier than scheduled, he decided it would be a good idea to wake me up and (just) offer sex. As a nineteen-year-old when a coach with whom I was traveling to a competition decided it would be a good idea to book a single hotel room with a king-size bed. As a child, an adolescent, and an adult on subways and buses, which certain men think is a good place to find female bodies to rub up against. As I learned from Criado Perez in *Invisible Women*, this is a thing the world over.

"Running while female" is also a thing. The last time for me was just a few years ago when I was about a mile from home and a man jumped out from the bushes as I was passing the spot where he'd been hiding. I was lucky that he wanted first to tell me about his big penis and wouldn't I want to see and touch it. It took everything I had not to make a sudden move but instead slowly, hopefully imperceptibly, to dial up the pace. It was when I heard his labored breathing and was about at my street that I took off. I no longer train at night or on trails and back country roads, even though these were always the best runs.

Beyond such jarring individual incidents there's the daily gauntlet of male aggression. These are probably physically safe moments, but you're never sure. The catcalls on the street: "I know you want a piece of this!" The higher-ups in white-shoe spaces who mouth a kiss or who touch you in a way that might be just friendly or else an intentionally ambiguous test. Rarely do these men and boys know much about you. Their interest is in your female form. Again, I want to be crystal clear: I'm not comparing myself to women who've been raped or are victims of domestic violence. Unwanted penetration by a penis is categorically different from what I've experienced, as are physical beatings by an intimate partner—which is in part why both are their own legal category. My point is you can be lucky not to have been raped, lucky to have partners for whom these acts are deeply anathema, lucky to live in places that aren't lawless, and still suffer a series of lesser violations that begin in childhood and persist through most of your life. It's part of a female's everyday anticipation of what's safe and possible. Among gendered experiences, the nature and ubiquity of this one hardwires your brain.

Sexual and domestic violence obviously cause individual harm, but there's collective harm too. The World Health Organization has called it "a major public health problem." Victims may die or suffer serious physical injuries and experience "depression, post-traumatic stress and other anxiety disorders, sleep difficulties, eating disorders, and suicide attempts." Children "who grow up in families where there is violence may suffer a range of behavioral and emotional disturbances" that "can also be associated with perpetrating or experiencing violence later in life." Society loses productivity as a result of victims' "isolation, inability to work, loss of wages, lack of participation in regular activities and limited ability to care for themselves and their children." Where legal and medical systems are in place and functional—which can't be assumed, even in parts of the United States—society also loses in the form of costs to these institutions.

Sex-based violence is an old, entrenched problem but it's not inevitable, at least not in its present magnitude. We know that it's propped up by a combination of informal social norms and laws that are either permissive or simply inadequate. The most obvious are criminal and civil laws and the apparatuses that are designed to deter, punish, and compensate for sexual violence. Beyond whatever's going on in Botswana, I'm thinking about California state court judge Aaron Persky, who, in 2016, sentenced Stanford swimmer Brock Turner to six months in the county jail for sexually assaulting an unconscious woman. Persky justified this leniency on the grounds that more time would have a "severe impact" on *Turner's* otherwise promising life. The law becomes outright encouraging in circumstances where the police are in on the action. As I was writing in 2023, two separate stories broke, one about a sheriff in Mississippi and another about members of the Iranian Revolutionary Guard. These perpetrators made a habit of taking sexual advantage of their female prisoners while others in law enforcement either looked the other way or helped to cover up their crimes.

During Covid, I watched telenovelas to brush up on my rudimentary Spanish—and, yes, because they're addictive. It's clear that current productions are intentionally undermining the informal norms that facilitate sexual and domestic violence and *femicide*. (Naming it as such rather than defaulting to the generic, male-patterned *homicide* is a part of that effort.) The telenovelas have strong female characters who have taken on sexual and domestic violence as their fictional cause—either toward law reform ends or for good old *venganza*—and all include male allies who are undoubtedly masculine but without the toxicity. We do ourselves a real disservice when we assume that male libido + strength = rape. This isn't the only dumb, destructive equation humans have come up with over the millennia, but it's on the short list.

Caregiving Norms

Unlike sexual violence, caregiving is a good and necessary thing, whether we're talking about rearing kids, tending to ill and disabled adults, or helping the elderly. Because we're all physically helpless and vulnerable at points in our lives, every one of us need caregivers, and we all benefit from norms or laws that support them. The question isn't whether caregiving should exist but rather what counts as such, who should do the work, how it should be done and valued, and the structures that are in place to support it.

Caregiving is both direct and indirect. A full-time working parent is a direct caregiver when they're not at work and "hands-on" with a child, and they're an indirect caregiver when they're at work earning money to provide for the family and hands-on care is done by someone else. Both kinds of caregiving are essential, but I'm focused here on direct caregiving.

Over time, in both patriarchal and matriarchal societies, females have done the vast majority of hands-on caregiving, and this pattern continues even in modern, egalitarian societies. In the United States, for example, about 75 percent of adult caregivers—those who work with the elderly and physically and mentally disabled adults—are female. According to DataUSA, which makes the government's statistics publicly accessible, the overwhelming majority of those who work with children—more than 90 percent—are female. Things are changing some at home, but as Julia Haines of *U.S. News & World Report* put it in advance of Mother's Day in 2023,

> It's a deeply rooted truth in the U.S.: Women carry a heavier child care burden than men. Gender gaps related to child care impact working women, stay-at-home moms and single parents alike, with stubborn societal norms and the high costs associated with child care leaving little room for improvement.

In opposite-sex couples, 95 percent of stay-at-home "parents" were mothers. Seventy-five percent of single "parents" are mothers—and that share jumps to 90 percent in the District of Columbia and 80 percent in Mississippi, Rhode Island, and South Carolina. The most progress toward equalizing childcare responsibilities has been among couples who are both working full-time: both are spending between 1.5 and 2 hours per day on childcare.

The main difference between patriarchal and matriarchal societies is or was the nature of and value attached to caregiving work. In matriarchal societies, the traditional maternal role is paired with political power because it understands and treasures women as life-givers and as guardians of the culture and cultural transmission. In patriarchal societies, women are valued for their childrearing role but not with political, cultural, or financial capital. The latter is a particular problem for caregivers in general, as much of this work is unpaid. In *Invisible Women*, Caroline Criado Perez reports, "Globally women do three times the amount of unpaid care work men do; according to the IMF [International Monetary Fund], this can be further subdivided into twice as much childcare and four times as much housework."

This sexed division of labor, as well as the tendency for it to be unpaid or underpaid, exists because, as compared to males, females as a group are assumed to be predisposed to caregiving by their natures. It's also assumed that this is a reasonable state of affairs given our two physical forms—that is, because of sex differences in human reproduction and physical performance. Given the facts of male and female biology, the extent to which the "groundwork" of caregiving is assigned to females across human societies, and that our animal kin mostly follow the same patterns, it's hard to make the case that there isn't a biological component to the role allocation.

The problem lies in the policies, structures, and norms that simultaneously deprive us of choice, devalue hands-on caregiving in relation to other work, discount or ignore the possibilities for males who would also be good caregivers, and fail to support us when we fulfill our roles.

Females are deprived of choice by norms that leave us the last person standing whose innate drive or inculcated sense of responsibility or both make abandoning a vulnerable family member, especially a young child, a near impossibility. There's a reason most single parents are women and that it's seen as "unnatural" not just "wrong" when mothers leave.

A devastating article by Patrick Strickland published by Al Jazeera around the time of the 2016 election told the story of Katie Bennett, "a 30-year-old single mother of three" who lived in a "rundown trailer" at a distance of "a few kilometers" from Booneville, Kentucky. Bennett was "one of the 52 percent of Owsley County residents who rely on welfare and food stamps to survive, drawing $649 a month to feed her children and herself." Strickland wrote, "Without a car or an education, she has been unable to land a job and cannot easily make it to town to search for work." As a result,

> Like others with no opportunities and little hope, she has struggled to put enough food on the table for her children as food stamps have been cut time and again in recent years. Single mothers, she says, feel the cuts more than anyone else. "[Food stamps] don't last the whole month. It really does [scare] me. I raise my kids alone since they been born. It's harder for a parent to do anything now." The food stamps last "probably till the middle of each month," [she] says. For the rest of the month, she relies on the goodwill of neighbors and relatives to feed her daughter and two sons. Often faced with choosing between feeding herself or her children, [she] frequently goes to bed hungry. "The grown-ups can do without [food] more than kids. I love my kids; I want the best for my kids," she concludes.

There is no sense in which Bennett can be described as valued or adequately supported in her caregiving role. She is the archetype of the single mother living in poverty whose own prospects are hard to discern and whose children will inevitably struggle to live better lives given her circumstances. There's an almost complete mismatch between our stated interests in replacement-rate fertility and the sacredness of each child's life, on the one hand, and what we're willing to do for women who make this collective contribution, on the other.

That the problematic aspects of the caregiving story are sociological phenomena means we can address them. They're not baked into our DNA or otherwise inevitable. Some of it is just being concerned enough to devote financial resources to the problem. Some of it takes creativity, including rethinking workplace norms and structures to take female patterns—and evolving male patterns—into account. Even bringing childcare back into the workplace.

I say "bringing back" because in the past, of course, children were raised alongside their adults who cared for them as they labored. Sexed childrearing after the tender years—the term for the developmental period to about age seven—existed as it did because from that point forward, kids were little adults in training and our domains were themselves sexed. But also, company towns often had on-site childcare. In fact, it's this company-town model that has interested innovators like Carmi Medoff, who is building a business called Onsite Kids which reimagines childcare as a workplace benefit. Medoff embraces the idea of childcare as a "public good" and she sees on-site childcare as one of a number of win-win propositions. As she explained when we spoke:

The lost productivity associated with women staying home to take care of kids is an annual $122 billion loss to the economy. It's also a loss to individual employers who rely on workers who have kids to get their business done, and to their employees who could really use the paycheck and the nonmonetary benefits associated with relationships outside of the home but just don't have options for quality, affordable childcare. This combination is why the people who are most interested in the possibilities for on-site childcare as a workplace benefit tend to be business owners, often men, and working women in frontline industries such as manufacturing, retail distribution, health care, and hospitality.

I asked Medoff why we're talking about this now, and she responded that although it's been a problem for a long time, "Covid put it in the spotlight as we needed human-centric operations to go on and many simply stopped functioning" because our childcare model shut down and there was no backup plan. She added that there are "purists" on the left "who don't want for-profits anywhere near this sector" and "who decry the health care dystopia and don't want to see the model extended to childcare." The concerns on the right are the classic ones, that "children belong at home with their moms and that the government shouldn't be involved with family matters." But there's bipartisan interest anyway because it's a commonsense solution to a significant public dilemma.

The Mental Health Epidemics Associated with Female Puberty

A headline in the *Atlantic* in February 2023 read "America's Teenage Girls Are Not Okay"—they "are 'engulfed' in historic rates of anxiety and sadness," went the lede. Another headline in the *Washington Post* on August 8, 2022, read "Why Tween Girls Especially Are Struggling So Much." And then there was this, in the *Wall Street Journal* on February 13, 2023: "Teen Girls Experiencing Record Levels of Sadness and Suicide Risk, CDC Says."

Adolescent girls as a group are in crisis. Although they've always had higher rates of mood disorders, the Centers for Disease Control and Prevention (CDC) reports that they're currently experiencing skyrocketing levels of anxiety, depression, persistent sadness, and hopelessness. In some cases, these rates are double and even triple that of males in the same age cohorts: high school girls in the United States today are three times as likely to suffer a major depressive episode as high school boys. Up to

57 percent of females in high school, compared with 29 percent of males, are likely to feel hopelessness. That's an astonishing 28-point difference.

What's most alarming is that this number represents an almost 60 percent increase from 2011, when 36 percent of high school girls and 21 percent of high school boys reported feeling hopeless. In 2021, 30 percent of high school girls seriously considered suicide, compared with 14 percent of high school boys. (These numbers were up from 19 and 13 percent in 2011.) The rate of attempted suicide was also nearly double: 13 percent of high school girls compared with 7 percent of boys attempted suicide (up from 10 and 6 percent). The pandemic contributed to increases in this period, as did a new willingness to report on mental health issues, but the skyrocketing numbers and rising disparities among girls and boys started much earlier. We tend to read more stories about high school students, and so the fact that "emergency-room admissions for self-harm among 10-to-14-year-olds tripled between 2009 and 2015" is especially shocking.

In a piece written for the *Washington Post*, psychologist Jelena Kecmanovic explained "the perfect storm" that's hitting girls in this age cohort. The sociological or constructed causes of these increases haven't been disentangled, but the candidates include "climate change and social upheaval"—which Kecmanovic explains "aren't just abstractions" for kids growing up today, it's "their future"—to the neurological effects of seemingly inescapable technologies including the blue light (from computers and devices) and social media. Both boys and girls are exposed to all of this, of course, but just as sex hormones impact the male and female brain differently, so do these technologies. As I mentioned in chapter 1 in connection with the introduction to brain sex differences, boys are more likely to play computer games and girls are more likely to spend time on social media. The latter "can be a wonderful source of support" but, unlike gaming, it can also be "a crushing blow" to their "self-esteem and psychological well-being."

Responsible reporting suggests that the social media story—the causal contribution of Instagram, TikTok, and other platforms to the epidemic—is complicated. But I don't know anyone who understands the female brain and how the technology works who isn't doing what they can to keep their girls off of it. In other words, they're not waiting for conclusive evidence of a causal connection to keep their girls out of what is effectively an unregulated social science experiment; they're taking a precautionary approach.

A precautionary approach was urged for all of us by no less than the surgeon general of the United States, Vivek Murthy. In May 2023, Murthy issued an official advisory—a "public statement that calls the American people's attention to an urgent public health issue"—warning parents to limit their children's use of social media because of a "profound risk of harm to the[ir] mental health and well-being," explaining, "We do not yet have enough evidence to determine if social media is sufficiently safe for children and adolescents." In an interview with the *New York Times*, Murthy emphasized that adolescents "are not just smaller adults," they're "in a critical phase of brain development." What the *Times* story didn't mention is that, consistent with sex differences in brain development, the effects are different for girls and boys. Per the advisory:

- "Adolescent social media use is predictive of a subsequent decrease in life satisfaction for certain developmental stages including for girls 11–13 years old and boys 14–15 years old."

- Among fourteen-year-olds, "greater social media use predicted poor sleep, online harassment, poor body image, low self-esteem, and higher depressive symptom scores with a larger association for girls than boys."

- "Social media may also perpetuate body dissatisfaction, disordered eating behaviors, social comparison, and low self-esteem, especially among adolescent girls."

These details are important, and so it doesn't make sense to erase sex from the story and to write only generally about "children" and "adolescents," because the developing brain, puberty, and the age of onset of puberty are different in boys and girls. It's always been lower—by about two years—in girls, and it's gradually dropped over time for both. But recently, the age of onset of female puberty has dropped more sharply than the age of onset of male puberty, and so the gap has grown.

The first sign of female puberty—breast budding or *thelarche*—used to be at about age eleven, followed by the first period—*menarche*—at about age thirteen. But "by the two-thousands, new research had found that eighteen percent of white girls, thirty-one percent of Hispanic girls, and forty-three percent of Black girls had entered thelarche by age eight." That's in the third grade, well before the start of middle school depending on the school district!

The causes of this recent drop also remain to be disentangled, but according to Jessica Winter, reporting for the *New Yorker*, they are thought to include "increasing rates of obesity; greater exposure to endocrine-disrupting chemicals found in food, plastics, and personal-care products; and stressful or abusive home environments." Given the race disparities in the U.S. data, also on the list of causal contributors are "structural racism and failures of environmental justice." These last two aren't just tossed in by advocates: it's a fact that the onset of puberty is sensitive to stress levels in other contexts and levels of endocrine disruptors in the environment are tied to race.

Early puberty means the girl is subjected to the cyclical onslaught of hormones in a different period of brain development, not only disrupting that process (which involves learning social and emotional skills including self-regulation), but also adding neurological stressors, especially anxiety and depression—the very conditions that describe the present epidemic—to the mix. Of course, it also means that she'll be dealing with breasts and fat development as well as her period when she's still just a kid. As Winter put it,

> early puberty presents something of a physics problem—how do we measure the passage of time? The bone X-ray may best illustrate the dilemma: a medical assessment that assigns the child to a skeleton that is older than the child herself. A tall, developed ten-year-old who has reached menarche may not be chronologically older than a petite, flat-chested ten-year-old who has not—but she is, in a real sense, physically and even experientially older. Adults and other children will almost inevitably relate to the girl differently—and not necessarily even in a sexualized way, although that is of grave concern; but intellectually, socially, emotionally. They may have advanced expectations of her, and she may strive to meet those expectations or fail to, and, either way, that cycle of stimulus and response is determining her place in her social milieu, conjuring a mirror in which she sees herself, and wiring her brain in configurations that subtly differ from those of her average-developing peers. Nature begets nurture. For this girl, the hands of the clock simply go faster.

Early puberty has both mid-term and long-term effects. The mid-term effects include "earlier onset of sexual activity, higher number of sexual partners, and higher likelihood of substance use, delinquency, and low academic achievement" as well as "depression, anxiety, eating disorders, and antisocial behaviors." The long-term effects include a "higher risk for obesity, type-2 diabetes, breast cancer, and heart disease." Given these risks, a parent and the child's pediatrician might responsibly choose to use puberty blockers to delay its onset.

Public health authorities sounding the alarm on the epidemic more generally make clear that it isn't only causing significant individual harm; it also has an expensive set of collective effects. That suspected causal contributors—technology and endocrine disruptors, for example—are human-made means that we can work to address them.

Men and Boys

Most of us have grown up in patriarchal or male-dominated societies that have been this way for centuries. In the United States, it's only been fifty years since the women's liberation movement managed to tear down the formal barriers to higher education and the workplace. It's a cliché but true that we've come a long way and that there's still lots to do. Even as the gaps are closing, men on average earn 18 percent more than women; households headed by men have 45 percent more wealth than households headed by women; men are still overrepresented in positions of power in government, private industry, and the professions; and there are still swaths of high-value opportunities from tech jobs to the arts that largely remain in male hands.

Behind these numbers, though, outside of elite circles where most people live, it's clear that the male half of the population is struggling. This isn't just a conservative talking point about the progressive assault on masculinity, although it is also that. Writers including Susan Faludi (*Stiffed*) and Hanna Rosin (*The End of Men*) warned us years ago that things were heading in this direction. More recently, Richard V. Reeves of the Brookings Institution laid out the case starkly in his also-essential book, *Of Boys and Men: Why the Modern Male Is Struggling, Why It Matters, and What to Do about It* (2022). In it, Reeves writes of the three-part "male malaise" that should concern us all: Dropping rates of participation and achievement in education. Dropping rates of participation in the labor force. Dropping levels of engagement with women, families, and friends.

On education, Reeves puts the matter bluntly: "The educational underperformance of [the male] half the population is now a routine fact to social scientists, one to be added to the standard battery of statistical controls." In the United States, this is true of boys from kindergarten all the way through higher education. The data are truly disturbing:

- "Girls are 14 percentage points more likely than boys to be 'school ready' at age 5," even "controlling for parental characteristics."

- A "6-percentage-point gender gap in reading proficiency in fourth grade widens to a 11-percentage point gap by the end of eighth grade."

- "In math, a 6-point gap favoring boys in fourth grade has shrunk to a 1-point gap by eighth grade."

Beyond secondary school, young men are now less likely than young women to go to college, and in college, "taking into account other factors, such as test scores, family income, and high school grades, male students are at a higher risk of dropping out" than "*any other group*, including poor students, Black students, or foreign-born students." Women now earn 57 percent of bachelor's degrees.

The causal contributors to these sex-based disparities in educational achievement include a combination of nature, educational policies and practices, and boys' and men's sense of their value and prospects. Reeves emphasizes how important it is for policymakers not to ignore nature and specifically sex differences in brain development. Here, he keys on "the biggest difference" between males and females: it's "not in *how* female and male brains develop, but *when*"—and on the fact that "the relationship between chronological age and developmental age is very different for girls and boys." He adds that this developmental difference is most important in high school, when the "gender gap in the development of skills and traits most important for academic success is widest" and also "precisely the time when students need to be worrying about their GPA, getting ready for tests, and staying out of trouble."

His conclusion is unavoidable: "From a neuro-scientific perspective, the educational system is tilted in favor of girls."

On the labor market, the message is that as the economy becomes more skill-based, men are dropping out. Writing in Quillette in 2021, the data scientist and criminologist Vincent Harinam warned that, "[b]ased on the linear trendline, the male labor force participation rate will continue to decline, falling below 65 percent for the first time by 2040." That means *more than 35 percent of men of working age will be out of the workforce.*

Nicholas Eberstadt of the American Enterprise Institute is probably the leading expert on this phenomenon. His book *Men Without Work: America's Invisible Crisis* (2016) is a must-read for anyone who wants to dig into the details. As he wrote in the *Wall Street Journal*, "During the past half-century, work rates for U.S. males spiraled relentlessly downward. America is now home to a vast army of jobless men who are no longer even looking for work—roughly seven million of them age 25 to 54, the traditional prime of working life." Reeves adds that already before the pandemic, a whopping "[o]ne-third of men with only a high school education [were] not in the labor force."

Prime-age male "unworkers"—Eberstadt's term for this cohort—aren't just out of the labor force; they're also not doing "much that's constructive" with their time otherwise: "Most are neither looking for work nor trying to develop skills that would make them employable. They're not 'helping around the house' or 'caring for others' or 'volunteering and engaging in religious activities.'" Most, Eberstadt concludes, are on screens.

Why is this happening? Reeves cites a "one-two punch, of automation and free trade." First, fewer jobs—only "one in ten"—still require what he calls "heavy work" and the "shift away from jobs requiring physical strength is going to continue." Further, nine out of ten jobs are subject to automation. Broken down by sector, this includes "70 percent of production jobs, 80 percent of transportation jobs, and 90 percent of construction and installation jobs." And men "often lack the skills necessary in an automating world." By contrast, women tend to be in "automation-safe occupations, such as health care, personal services, and education." Then there's free trade:

> [T]he political elite spent decades complacently arguing that on net, and in the long run, free trade is good. And so it is. By definition, however, this means that some people, in some places, are being hurt right now. Not much was done to help these people, even by center-left politicians who claimed to be on the side of the working class. The victims were basically left behind, told to buck up their ideas, engage in some "lifelong learning," and get with the program.

Reeves and Eberstadt are both clear that the costs of what Eberstadt calls the "new normal" for the society are extraordinarily high. My male friends and students who have lived in the environments in which this demographic is most prevalent add that these are not just numbers churned out by experts at fancy East Coast think tanks. On the ground, in the real world, the fear is palpable.

A friend who grew up in a former auto town in Michigan describes himself as having been one of a handful of black boys whose parents had enough money to send their kids to the local private (almost all-white) high school. Failure wasn't supposed to happen to them, but it did. One died by suicide. Another is in prison. A couple of others have struggled continuously since high school. When I asked him why he thinks this happened, he explained:

> We all went to high school planning for college. But that doesn't always work out for everyone, and that's fine. Or at least it used to be.

Before, if it turned out you weren't going to college after all, you were still able to graduate high school with optimism for your life because you knew there was meaningful work, a job that made you part of the community, and which would allow you to provide for your family, for your kids. And here's the thing, the jobs were good jobs—they afforded you a middle-class life. If both you and your spouse worked for Ford, you were a factory worker who could buy a house with a pool. Then all of a sudden the jobs just disappeared.

He paused before saying, "And nothing replaced them." This hit everyone hard, he added, "but it hit boys and men harder because the jobs that disappeared were the ones men especially counted on. The physical ones, the ones where you work with your hands." There were drinks involved in our conversation and perhaps that allowed for what might otherwise have been a verboten add in our academic circles: "It wasn't just that these were jobs men could do because they're male," he said. "It was also that because they're male it's important that there were physical jobs."

Those jobs and the life they afforded with a family and kids were all part of what it meant to know you had a future. One leads to another and then all of a sudden it doesn't. That leaves boys and young men without hope and that's not only devastating for them, it's dangerous for society.

My friend is convinced that he got out because his parents had gone to grad school. They were professionals, and their own world hadn't come crashing down when the auto industry relocated. His family occupied the same geography as the displaced but they otherwise lived different lives and so his prospects were on firmer ground.

It should go without saying that men who are, or are perceived to be, unsuccessful in education and the workforce are also struggling with relationships. However we might rationalize our loneliness, humans are at bottom a social species, which is in part what makes these stats from a recent article in the *Hill* by Daniel de Visé so upsetting:

- "More than 60 percent of young men are single, nearly twice the rate of unattached young women." "Only half of single men are actively seeking relationships or even casual dates, according to Pew. That figure is declining."

- "Men in their 20s are more likely than women in their 20s to be romantically uninvolved, sexually dormant, friendless and lonely."

- "Fifteen percent of men report having no close friendships, a fivefold increase from 1990, according to research by the Survey Center on American Life."

One of the things that struck me as I was researching this topic is that, unlike men, women tend to want to partner or marry "up" or at least "over" educationally and economically. Evolutionary biologists note the similarity to female preferences in the animal kingdom. It may have originally been adaptive, of a piece with the blow to self-esteem many men experience when they have a female partner who makes more money than they do. Understanding that they may be biologically based, it's worth considering whether both tendencies aren't today something of a self-defeating proposition.

Like the costs of men's educational and workplace losses, those associated with their lost relationships are also high. We know that health-wise, in general, "People with a strong social network are less stressed, more resilient and more optimistic. They're more likely to be a healthy weight and less likely to suffer cognitive decline. They also enjoy some protection from cancer, heart disease and depression." One study found that "long-term social isolation can increase a person's risk of premature death by as much as 32 per cent."

Finally, men are more likely to suffer deaths of despair—those that result from chronic liver disease, drug overdoses, and suicide. According to the National Center for Health Statistics, "for every 100 females between the ages of 15-24 who died by suicide, 400 males died. For every 100 women over the age of 75 who died, 930 men died by suicide."

The label "deaths of despair" has primarily been attached to white Americans. Because of this, I was struck when I read that "midlife mortality rates" are "far higher" among American Indians and Alaska Native people who had been left out of earlier analyses. Beyond the fact that this was yet another harmful erasure, I couldn't help but tie the data to the crazy rate of sexual assault and domestic violence in Indian Country. It has to be that to address that epidemic (and so to do right by women and girls) you have to address the high rates of despair (and so to do right by boys and men). It also has to be that targeted—sex-smart—solutions are more likely to be effective than sex-blind ones. Indeed, according to Everett Rhoades, the first Native director of the Indian Health Service, for men, "Programs targeted to anomie, loss of traditional male roles, and violence and alcoholism are among the most urgently needed." Also, it wouldn't make sense to require program directors to teach—or women and girls to attend—a coed session on anomie-driven sexual violence.

The first wave of conscriptions in early 2022 triggered protests across Russia and young men fled the country in droves. I met one of them in Morocco. Nikolai didn't want to be part of a senseless war. People who watch only state-run television in his country believe the story about a limited "special military operation" to protect Russia from Ukrainian

Nazis, but no one else does, he said. He didn't want to kill young Ukrainian men "just like me" and he didn't want to die on the front line, where untrained Russian troops are sent as cannon fodder. The senselessness of it was clear as we talked. Nikolai felt both fortunate and guilty that he had the means to leave when most of his compatriots didn't, and it was really hard to walk away from his mother—a patriot who believed her son should do his duty by the motherland. She'd been misled by the government's propaganda, but he could neither convince her of this nor disappoint her in such an existential way. But then, he said, as the casualties mounted and the real news managed to penetrate her barriers, "some kind of wonderful happened." She told him to go.

There's no end in sight for the war in Ukraine. Because the Russian military needs conscripts, and to deal with draft dodgers like Nikolai, President Vladimir Putin adopted a new law in April 2023. By its terms, however they move, wherever they turn, young men will find themselves trapped: They won't be able to claim they didn't get their draft notice because the law treats it as having been received a week after it lands in their government email inbox—even if they never signed up for an account on the site. Those who are drafted are prohibited from leaving the country. Being a no-show can trigger "a driver's license suspension, along with a block on registering real estate and other property and on receiving a bank loan."

There's no end in sight either for women and girls in the United States on the abortion front, where laws that would box them in are also proliferating. As I write, the legislature of my home state of North Carolina has just passed a twelve-week abortion law, in part—its proponents say—to signal to out-of-staters that we will not be a sanctuary when their own jurisdictions have shut down access. Twelve weeks is three months, the first trimester of pregnancy, and that can sound and even be reasonable in certain circumstances. Nine out of ten abortions in the United States already take place in that period, and it's certainly better than the six-week law passed by South Carolina. But the details in the North Carolina law are torturously complex: No medical abortions—which are the majority of those that take place in the state—after ten weeks. There are numerous attestation requirements and then three in-person appointments, days apart, before an abortion can take place even within the allowed period. There are new regulatory and licensing requirements for clinics that offer abortions, which inevitably means there will be fewer of them. Some of the bill's sponsors acknowledged that these are just interim compromises on the way to a total ban.

Part III

On Sex
and
Gender

7

The (Un)Lawfulness of Regulating on the Basis of Sex and Gender

RUTH BADER GINSBURG WAS "famously tiny." This extended even to the sound of her voice. But as Sharron Frontiero recalled it years after Ginsburg had laid the foundation for modern sex discrimination law partly through Frontiero's case, "the silence of her was like an engine at the middle of the universe." To understand Ginsburg's revolutionary contributions—and the quiet, methodical way she made them—it's best to start with her favorite part of the Constitution: Section 1 of the Fourteenth Amendment (1868).

The Fourteenth Amendment is one of a set that was ratified after the Civil War to reunite the country and to secure national-level rights for the previously enslaved population, rights that the states couldn't deny. It's not an understatement to say that much of what it means to be "American" today is embedded in the short, powerful text that is

Section 1 of the Fourteenth Amendment. Read it slowly to see how this is true:

> No State shall make or enforce any law which shall abridge the privileges or immunities of citizens of the United States; nor shall any State deprive any person of life, liberty, or property, without due process of law; nor deny to any person within its jurisdiction the equal protection of the laws.

The *Privileges or Immunities Clause* is the one Myra Bradwell and Matthew Carpenter used (unsuccessfully) in 1873 to argue that national-level citizenship includes the right to practice one's profession regardless of sex.

The second provision is the *Due Process Clause*. Beyond its guarantee of fair process, usually a hearing, since the early twentieth century its embedded word *liberty* has been read to include the substantive rights to procreate, to marry, and to raise children according to one's own values and ideals. As we saw in chapter 6, for fifty years from *Roe* to *Dobbs*, *liberty* also included a woman's right to choose to terminate a first-trimester pregnancy. Abortion "resided" here.

The third provision is the *Equal Protection Clause*. When it was adopted three years after the end of the Civil War, the animating idea was that black people would no longer be subject to "slave codes"—the completely different set of laws that, for about three centuries, had applied to bonded people of African blood. Instead, at least as a matter of law, they would be considered the same as white people. That this idea was too revolutionary for the time is why Reconstruction failed, why "Jim Crow" and its "separate but equal" doctrine prevailed for over fifty years, and why we're still working on racial equality today. Throughout the Reconstruction era, however, the clause continued to promise "any person" the equal protection of the laws.

It's obvious to us today that this promise applies also to sex. But it wasn't always so. Early on, there were lots of women for whom rights were paramount who didn't consider themselves to be looking for equality or equal treatment in relation to men. They sought liberty, or simply the opportunity to get things done. Historically, equality was tightly bound up with race, and the word *equal* meant "the same as." This was something society wasn't prepared to say about sex.

The most important advances in women's rights in the nineteenth century occurred without recourse to equal protection law. Family law abandoned the merger doctrine, which conceived of a husband and wife as one, with the husband being the public-facing side of the couple and the wife the private-facing side. Contracts law abandoned the doctrine that prohibited married women from making deals, even on their own behalf. Property law stopped subscribing to the rule that automatically gave a husband his wife's property on the occasion of their marriage. In hindsight, it's possible to see these reforms as having been equality-minded, but at the time they were mostly treated as practical solutions to new problems created by industrialization, the mass movement to cities, and even to market fluctuations.

For example, in the mid-nineteenth century, ownership of the mansion in Macon, Georgia, in which my enslaved ancestor Ellen Craft lived as a house servant was transferred from her enslaver, Robert Collins, to his wife, Eliza—Ellen's white half sister—probably to avoid his creditors. To these same asset protection ends, Eliza formally owned other business properties in town that in fact belonged to Robert or to her father, James Smith, who was also Ellen's father. Ilyon Woo's beautiful biography of Ellen and her husband, William, *Master Slave Husband Wife: An Epic Journey from Slavery to Freedom* (2023), is chock-full of such nineteenth-century history.

My favorite example, though, is from the early twentieth century, when the New York State Woman Suffrage Association argued that women should be given the vote so that they could do their assigned work both inside and outside the home. They would be better stewards of "health and welfare" if they could also focus on "municipal house-cleaning." But women who opposed the Nineteenth Amendment, which was ratified in 1920 and gave women the right to vote, responded that their effectiveness came precisely from being "outside of politics" and that "[they] would lose immeasurably if this power were taken from [them]." The arguments were about expanding women's rights but the mindset and language had little to do with the "sameness" of males and females that "equality" suggested.

There were, of course, women's rights advocates in the nineteenth and early to mid-twentieth centuries who did focus on "sameness" and "equality." For example, Crystal Eastman, a cofounder of the ACLU in 1920, argued that the day when Tennessee first ratified the amendment was "a day to begin with, not a day to end with." Eastman's goal was freedom and independence for women, and she saw sex equality as necessary to those ultimate ends. She focused on the ways in which men and women are the same in relation to the things that matter once we dispense with the artificial barriers that say otherwise, barriers like men's feigned ignorance about and incapacity in the home, and the false notions that women can't compete in or should be protected from the dangers that lurk in the workplace. Whether it was putting in long hours in a factory, bartending, or sitting on a jury, women are like men so they should get equal treatment. But these advocates mostly hadn't been persuasive. People continued to think that it was rational for policymakers to take sex and sex role differences into account when legislating for society's health, safety, and morals.

As a result, through the mid-twentieth century, sex classifications—laws that treat men and women differently—remained ubiquitous, and the Court had never used the Equal Protection Clause to invalidate

Mexican-born American sculptor Robert Graham (1938–2008) is known for his bronze renditions of the human form. He created the dual figure *Olympic Gateway* for the 1984 Games in Los Angeles. The original sits at the entrance to Los Angeles Memorial Coliseum. Graham made this copy for the entrance to the International Olympic Committee's headquarters on the shores of Lake Geneva in Lausanne, Switzerland. The work updates the Greek and Roman marbles of the athletic male form in competition, adding the athletic female form and presenting both in the immediate aftermath of competition. A wonderful book by Graham, *Studies for the Olympic Gateway* (1984), describes his artistic process from concept to drawings to clay casts to the bronzes, and the Olympians—a male water polo player and a female long jumper—who were his models. *Doriane Coleman with permission from the Robert Graham Studio*

The book's front endpapers feature an extraordinary set of mosaics found at the Villa Romana del Casale in Sicily. The work of North African artists in the fourth century, their seeming modernity has fascinated visitors since they were discovered by archaeologists in the early 20th century. They are sometimes called the "bikini girls" because their breastbands and loincloths resemble two-piece bathing suits, but they're perhaps better compared to today's elite female track-and-field athletes. The twist is that in the classical period, male athletes competed in the nude—as Graham's *Olympic Gateway* has it—to honor the gods with a display of the beauty of the perfected body, whereas female athletes were partially clothed—as the mosaics reflect—with a form of what we would call sports bras (for the same reasons we wear these) and competition briefs (ditto). Today, by contrast, female athletes who compete this way are criticized for being underdressed and sexualized rather than liberated and beautiful, and it's suggested that their uniforms should cover more of their bodies. *Julian Money-Kyrle / Alamy*

Featured here as examples of this last provocation are Faith Kipyegon (Kenya) and Genzebe Dibaba (Ethiopia) at the conclusion of the women's 1,500-meters final at the 2016 Olympic Games in Rio de Janeiro, where Kipyegon took the gold and Dibaba the silver. Two years later, in 2018, Kipyegon gave birth to a daughter and then, in 2021, she repeated as Olympic champion at the Tokyo Games. In the aftermath of that victory, her visible pregnancy skin folds became the subject of a global conversation: Were they beautiful and inspiring as expressions of motherhood and of mothers as champions? Or were they ugly because her midsection was no longer smooth and ripped as it was in 2016? It seems that men and women had different reactions. In any event, as the rear endpapers detail, in 2023, when Kipyegon broke the world records at 1,500 meters (3:49.11) and the mile (4:07.64), her midsection was covered. *DPPI Media / Alamy*

Jarmila Kratochvilová set the still-standing world record in the 800 meters—1:53.28—in 1983 in a race in Münich, Germany, in which I also competed. Along with the women's world records at 100, 200, and 400 meters, the 800-meters world record is widely considered to have been the product of doping. In this photograph from the 1985 Bruce Jenner Classic in San Jose, California, Kratochvilová leads 1984 Olympic bronze medalist Kim Gallagher, Doriane Lambelet, and Gail Conway. I started law school that fall. *David Madison / Getty Images*

Caster Semenya of South Africa, featured in chapter 1, is shown here gapping the field of eight to win gold in the women's 800-meters final at the Rio Olympics in 2016. The others in order of finish were Francine Niyonsaba (Burundi) for the silver and Margaret Wambui (Kenya) for the bronze, followed by Melissa Bishop-Nriagu of Canada (4th), Joanna Jóźwik of Poland (5th), Lynsey Sharp of Great Britain (6th), Maryna Arzamasava of Belarus (7th), and Kate Grace of the United States (8th). I list all their names because in 2016, the eligibility criteria for "women's" events were suspended, allowing certain genetic males to compete in the female category with no conditions attached. *Ezra Shaw / Getty Images*

Myra Bradwell featured in chapter 2, was one of the first women in America to take and pass a bar exam. She is most famous today for being the plaintiff in *Bradwell v. Illinois* (1872), in which she sought to have the Supreme Court overturn her state's decision to deny her a license to practice law on the grounds that she was female. In her brief to the Court, she and her lawyer debuted the argument that the 14th Amendment bars unreasonable sex classifications just as it bars unreasonable race classifications. They were unsuccessful—it would take another hundred years before Pauli Murray and Ruth Bader Ginsburg developed a winning strategy to use the amendment to those ends. Bradwell was much more, however, than the plaintiff in that case. The daughter of abolitionists, she made the undoing of sexist laws her life's work—and in this she was enormously successful. *Library of Congress / Getty Images*

The Senate confirmation hearings on Ketanji Brown Jackson's nomination to be an associate justice on the United States Supreme Court in 2022 were disrupted on the day that Senator Marsha Blackburn asked the nominee, "Can you provide a definition for the word 'woman'?" The ripple effects of that question and their subsequent exchange, featured in chapter 3, have been profound. It is fitting and reparative that this photograph—of Jackson and her teenaged daughter Leila at a different point in the hearings—also marked the occasion. Taken by Sarabeth Maney of the New York Times, it went viral on Twitter (now X). As reported by Gina Cherelus in the Times, "The image resonated with thousands of users across social media, especially mothers, who found the moment between Judge Jackson and her daughter to be inspirational. People have called it 'mom goals in a photo,' and said that 'this look of pride in Justice Jackson's daughter's eye speak volumes about what this means for little girls of color.'" Sarabeth Maney / The New York Times / Redux

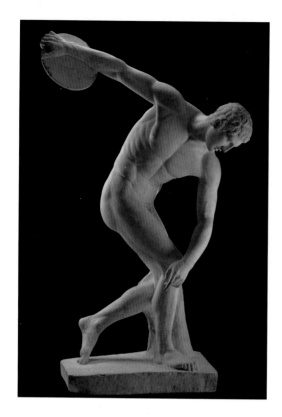

The Discobolus is believed originally to have been sculpted in bronze by Myron in the beginning of the classical period—around 450 BCE. It has survived only in marble copies, of which the second-century Roman "Townley Diskobolus" in the British Museum is one of the most famous. It is a representation of one kind of physical beauty: the harmonious development of body and mind.

British Museum Images

Olugbenro Ogunsemore's photograph of U.S. soccer player Oguchi Onyewu recalls the Discobolus and reflects the enduring fascination with the male form. The photograph was included in the inaugural *ESPN The Magazine* Body Issue (2009), which is perhaps best remembered for featuring Serena Williams on its cover. The issue was described in an ESPN press release as "a celebration and exploration of the athletic form" and "a testament to the work the athletes do, the effort they exert and the price they sometimes pay in reliance on their most important asset—their bodies." Olugbenro Ogunsemore

Featured in chapter 4, Michelangelo's *David* (1501-5) is probably the most famous sculpture of the male form. Charged by his Florentine patrons to represent the biblical king, he chose to do this by referencing the Greco-Roman classics. For a combative Florence, as art historian John Paoletti put it, the sculpture represented "the athletics of war." It was originally located in the Piazza della Signoria—the city state's seat of government—and then moved indoors in the 19th century to l'Accademia delle Arte. It remains there in a perpetual state of preservation. A full-size replica replaced it in the piazza. Over the centuries there's been ongoing debate about whether the sculpture is political, beautiful, or pornographic. Most recently, the latter claim came from a charter school in Florida that specializes in the classics. The commotion that resulted played into ongoing political battles over parental rights and the role of the public schools in matters of sex and gender. *Michael Gottschalk / Getty Images (in piazza) and Alberto Pizzoli / Getty Images (in rotunda)*

Here are two artistic representations of the female form from different periods and places. The first is an Indian sandstone from the 9th-10th century. It is impossible not to see it as a reference for one modern iteration of the ideal female form. The second is a torso of the goddess Venus dated to the Roman period. Described as "a delicate depiction of beauty, sexuality, and fertility" it is notable for having been sculpted out of gray-black marble—a material sourced in Africa and coveted by Roman emperors for its beauty and rarity. *Bridgeman Images (the black marble) and the Brooklyn Museum (the sandstone)*

The *Venus de Milo*, featured in chapter 4, is one of the most famous sculptures in the world. Found on the Aegean island of Milos in 1820 and attributed to Alexandros of Antioch between 150 and 125 BCE, it has been displayed at the Louvre almost continuously since 1821. As a representation of the ideal female form, the sculpture has been both revered and criticized. Throughout, two things have been clear: whether the representation is experienced as ideal or even beautiful is context dependent; but as artistic subject and fascination, the female body itself is enduring. *Hervé Lewandowski / Art Resource (profile headshot) and DEA Picture Library / Getty Images (full sculpture #122317493)*

The award-winning television show *Pose* is set in the 1980s and 1990s in New York City's house-ballroom world—the community of chosen families or "houses" formed by black and Latino LGBTQ people who were excluded from the white LGBTQ community and its drag scene during the height of the AIDS epidemic. For reference, Madonna's hit song "Vogue," containing the phrase "strike a pose," celebrates the competitions at these balls. This still, from one of my favorite episodes of the show, captures Michaela Jaé Rodriguez as Blanca Evangelista—the head of the House of Evangelista—and Indya Moore as Angel Evangelista on a day where the family was hoping to enjoy a day at the beach. In 2022, Rodriguez became the first trans woman to earn a Golden Globe for best actress in a television drama. In 2019, Moore, who is also a model and a civil rights advocate, was named by *Time* as one of the most influential people in the world. In announcing that designation, fellow trans woman Janet Mock described Moore as "the living embodiment of our wildest dreams finally come true."

Cate Campbell of Australia, featured in chapter 5, specializes in the 50- and 100-meters freestyle events. On her best days, she's the fastest female on the planet in the water. A five-time Olympian, four-time Olympic gold medalist, four-time world champion, and former world record holder, as this book goes to press, she is training for Paris 2024, which would be her sixth Olympic Games. She's long been active outside of the pool in matters of girls' and women's physical and mental health, driven by her own experiences with the male default in treatment and training and by the commitment she's made to use the privilege of her substantial platform to push for female-focused approaches. She has been especially thoughtful about how her work touches on trans women in sports. "We should acknowledge and celebrate both females and transgender women," Cate says. "In our push for equality, we've forgotten about diversity." And "[i]t's not a competition, it shouldn't be a competition, because we don't have the same physical experience." *Courtesy of TLA Worldwide*

Pauli Murray was one of the most important women in the history of equality law. Featured in chapter 7, she had a hand in designing the strategy underlying *Brown v. Board of Education* (1954) and *Reed v. Reed* (1971), the cases that laid the groundwork for modern approaches to unreasonable race and sex classifications. That you may not have heard of her, or only recently learned of her, says everything about the marginalizing effects of racism and sexism and nothing about the enormity of her contributions. Murray sent this photograph to Eleanor Roosevelt in the late 1950s, when she was thinking about how to apply the lessons from *Brown* to combat sex discrimination. In this she would pick up the baton from Myra Bradwell and hand it off to Ruth Bader Ginsburg. *Courtesy of the FDR Library*

Ruth Bader Ginsburg is widely hailed as the architect of modern sex discrimination law for her work as a lawyer in the 1960s and 1970s. That work resulted in pervasive law reform as the statute books—then chock full of old sex classifications—were rewritten to be the gender-neutral provisions we know today. This photograph was taken in 1977, three years before she was first appointed to the federal bench by President Jimmy Carter. *Courtesy of the United States Supreme Court*

Sandra Day O'Connor was the first woman appointed to the Supreme Court, by President Ronald Reagan in 1981. Ginsburg was the second, appointed by President Bill Clinton in 1993. Despite their political differences, Ginsburg and O'Connor were close, sharing a commitment to women's equality that played out in ways that set the stage for the current debates on sex and gender. This photograph of the two justices was taken at the Supreme Court by Michael O'Neill for Vanity Fair in 1998—two years after the Court's decision in *United States v. Virginia*, also featured in chapter 7. Michael O'Neill / Contour RA by Getty Images (RBG and SDO'C)

Mary Wollstonecraft featured in chapter 9, is sometimes described as the mother of modern feminism for her contribution to the centuries-long debate that is the Woman Question. In her book *A Vindication of the Rights of Woman* (1792), she advocated for girls to be educated alongside boys and against many of the social constructions that presumed that females—because of their physical forms and physiology—were less capable than males. In 2020, Maggi Hambling's *A Sculpture for Mary Wollstonecraft* was unveiled in a park in North London. It immediately became a sensation for its inclusion of a nude female. Hambling was unrepentant: it was for Wollstonecraft and about the conversation she ignited. Courtesy of the New York Public Library (painting) and Grim23 (statue)

one. The one big attempt to alter this dynamic was the Equal Rights Amendment (ERA), which Eastman coauthored. It was first introduced in Congress in 1923 and was never ratified despite its basic, seemingly innocuous language:

Section 1. Equality of rights under the law shall not be denied or abridged by the United States or by any State on account of sex.

Section 2. The Congress shall have the power to enforce, by appropriate legislation, the provisions of this article.

In general, sex-based policymaking was seen as supportive of women and of the social fabric of society, rather than as discriminatory (in the bad sense of the word). Forcing equality and sameness through the law was seen as antithetical to those interests.

By the 1950s, when Ruth Bader Ginsburg went to college at Cornell, enrolled in law school at Harvard, and then transferred to Columbia where she graduated at the top of her class, things were much better for women—including for exceptional women like her—than they had been in Myra Bradwell's day. That I can list these milestones proves the point. Skirts had made it to the Ivy League. Still, there was a long way to go.

Ginsburg's own experience—as captured by the movies that have been made about her life including *RBG* (2018) and the documentary *Ruth: Justice Ginsburg in Her Own Words* (2019)—embodies this mixed moment. In the mid-1950s, she was demoted by her employer when she became pregnant with her first child. At Harvard, the dean asked her and her female classmates why they were taking spots that belonged to men. Despite graduating from Columbia with the most stellar academic record, the judges and firms that normally hired the law school's top students wouldn't touch her. In 1960, Justice Felix Frankfurter denied her a Supreme Court clerkship because she was female.

That he had written the opinion upholding the reasonableness of a state law permitting women to be waitresses but not bartenders—on the view that the Constitution doesn't require states to keep up with "shifting social standards"—goes partway to explaining his views on female law clerks. But this was also the ethos of the time, and the law played its role in perpetuating it.

Things were different with race, at least on the books. The civil rights movement had been in gear for over a decade. Most importantly for our purposes, the legal case had already been made that discrimination on the basis of race is almost always wrong. Two Supreme Court decisions, *Korematsu v. United States* (1944) and *Brown v. Board of Education* (1954), paved the way for the eventual reforms.

In *Korematsu*, the Court held that the government's wartime internment of Americans of Japanese descent didn't violate equal protection, but it reached this abhorrent—since corrected—result only after deciding that classifications that

> curtail the civil rights of a single racial group are immediately suspect. That is not to say that all such restrictions are unconstitutional. It is to say that the courts must subject them to the most rigid scrutiny. Pressing public necessity may sometimes justify the existence of such restrictions; racial antagonism never can.

Today we refer to *Korematsu*'s "rigid scrutiny" requirement as the *strict scrutiny test*. It applies when measures differentiate among people who are members of a *suspect class*—"suspect" meaning that because of the group's shared characteristics, if the challenged measure curtails rights, there's reason to suspect that it's motivated by antagonism toward the group. There are three legally recognized suspect classes: race, national origin (including ethnicity), and alienage.

To show a violation of equal protection when strict scrutiny applies, the person complaining about unfair treatment has to prove that the measure differentiates among people on the basis of a suspect criterion or interferes with a fundamental right (like the free exercise of religion) *and* that they're part of the group being treated unfairly. The burden then shifts to the government to prove (1) that the measure is justified by a legally compelling interest—something that demands action—and (2) that the discrimination is necessary to achieve it. The key is that compared to the run-of-the-mill case in which legislative classifications are assumed to be *valid* (because legislators are generally rational actors) so that it's difficult to show a violation of equal protection, strict scrutiny review assumes they're *invalid* (because of the special stakes or the tendency not to be rational about race, ethnicity, or alienage) so that it's difficult for the government to justify them. The burden on the government is so steep that it's been said that strict scrutiny is strict in theory, fatal in fact.

Ten years after *Korematsu*, in *Brown v. Board of Education*, the Court held that racially segregated public education violates the Equal Protection Clause. The winning arguments against the "separate but equal" doctrine that had allowed segregation were textual and sociological: They focused on the "separate" side of the equation, showing that black children knew they were educated apart because of false notions about their inherent inferiority. This knowledge has lifelong, debilitating effects. As a result, the experience of the kids in white and black schools could never be equal, even if the facilities themselves were theoretically the same.

The movement for racial equality provided important perspective and inspiration for women's rights advocates. The ties between the two groups weren't always tight. There's great writing about their divisions including about how white feminists sidelined black women and their concerns, a subject I broached in chapter 3. Still, there were always important women who straddled the two groups.

One of these was the extraordinary Pauli Murray. Her book *States' Laws on Race and Color* (1950) was described by Thurgood Marshall as the "bible" for the civil rights movement, and the arguments she developed as a student at Howard Law School are said to have been crucial to the development of the winning strategy in *Brown*. In a 2021 documentary about her life, *My Name Is Pauli Murray*, Murray talks about resisting a world which saw "the Negro as inherently inferior to whites and women as inherently inferior to men." To capture this reflection, she called herself and other black women of the time "Jane Crow"— a moniker that the author of her comprehensive biography, Rosalind Rosenberg, chose for its title, *Jane Crow: The Life of Pauli Murray* (2017). Murray was dedicated—first as a lawyer, then as an educator, and finally as an Episcopalian priest—to taking down these intersecting hierarchies.

In 1961, her friend Eleanor Roosevelt ensured that Murray was included on President John F. Kennedy's bipartisan Commission on the Status of Women, of which Roosevelt herself was the chair. The commission's charge was to "examine discrimination against women and ways to eliminate it." Members "studied legislation and services that would help women more easily fulfill their roles, whether as housewives, workers, or citizens" as well as the set of laws that continued to subordinate them to men. In 1962, Murray wrote a memorandum for the commission titled "A Proposal to Reexamine the Applicability of the Fourteenth Amendment to State Laws and Practices Which Discriminate on the Basis of Sex Per Se." In form and method, it was a classic practitioner's memo. The issue it addressed was how to remove traces of sex inequality from the law. It proposed that everything—all of what Ginsburg would later call the "remedies" in the "arsenal"—should be on the table.

Murray's own preferred remedy was the Equal Protection Clause. As Ginsburg would later explain, "Pauli had the idea that we should interpret the text literally." *No State shall . . . deny to any person within its jurisdiction the equal protection of the laws.* Murray acknowledged the legal and political challenge this posed: no sex classification had ever been found to violate equal protection in part because, again, "equal" suggested "same," which wasn't a popular idea as applied to sex. But the clause had been used successfully to invalidate race classifications, and Murray—who was no stranger to the intersection of race and sex—believed that the two were a lot more alike than the law and other practitioners had yet recognized. Plus, she had the life experience, legal training, personal commitment, and work ethic necessary to build a real-world case against formal prejudice. Her ultimate goal was for everyone to be seen by the law as equally worthy of dignity and respect.

Nonetheless, Murray urged advocates not to concentrate on the "equality" aspect of equal protection law. Instead, she suggested they focus on the workaday lawyering necessary to prove—under the default rational basis test—that certain sex-based measures are unreasonable. "The central constitutional issue," she wrote, "is the *reasonableness* of each legislative classification based upon sex within the concept of equal protection of the laws." She added examples of laws that would make for good challenges, as she had done in her book *States' Laws on Race and Color.* She also described the sociological case against sex discrimination—its real-world, negative impact on women in general.

At the back of her memo, in view of an eventual argument to ratchet up the level of scrutiny that would apply to sex classifications, she included a chart detailing the analogy of sex to race. In that chart, which she captioned "Castelike Status of Women and Negroes," she included categories like "high social visibility" of race ("skin color and other 'racial' characteristics") and sex ("secondary sex characteristics"); attributes "ascribed" to race ("inferior intelligence, smaller brain, less convoluted, scarcity of geniuses") and sex ("ditto"); "rationalizations of status," "accommodation attitudes," "discriminations," and "similar problems" (followed by examples that were being used in race cases, with the word *ditto* for "sex").

By the time Murray's memo made it into Ginsburg's hands in the late 1960s, Murray was teaching at Brandeis University and no longer directly involved with the law; but she was on the board of the ACLU, which had earlier "identified [her memo] as a roadmap to begin combatting gender inequality." When Ginsburg joined the ACLU's effort as a volunteer lawyer, that road map forever bound the two women together. Indeed, based on its guidance, Ginsburg would soon take on *Reed v. Reed* (1971), the case that is famous in constitutional law for being the first in history to find that a state's sex classification violated the Equal Protection Clause and for laying the foundation for modern sex discrimination law. At a century's remove from Myra Bradwell, the Fourteenth Amendment, albeit a different provision than she and Matthew Carpenter had picked, would finally be read also to prohibit sexism in the law.

Sources sometimes describe Pauli Murray as an unsung hero. But Ruth Bader Ginsburg always recognized Murray's contributions. If you look at Ginsburg's submission to the Supreme Court on behalf of Sally Reed, you'll see Pauli Murray of Beacon Street in Boston credited as a coauthor. As Ginsburg put it in an interview in 2017, "We knew when we were writing that brief that we were standing on her shoulders." Ginsburg added, "We owe so much to [Murray's] courage, to her willingness to speak out when society was not prepared to listen." "Unlike Pauli who was way ahead of her time, I was there at the right time."

Different Treatment on the Basis of Sex as (Bad) Discrimination

Ruth Bader Ginsburg was ultimately committed to the proposition that women should be treated in law and in life as "full human personalities." But like others over the centuries who worked on the Woman Question, she wasn't blind to sex. She didn't pretend that sex doesn't exist or that males and females are the same or that sex itself is bad. Rather, she saw aspects of the law—chock-full as it was in her day with gendered artifice—as bad because they denied women the opportunity men had to develop, and society to benefit from their full human potential. Equal protection law as it had been developed in the context of race, and analogizing sex equality to race equality, were the remedies for that problem.

The first hurdle was getting the Court to agree with this proposition. *Reed v. Reed* (1971) was the vehicle. At issue in *Reed* was the constitutionality of an Idaho law that automatically designated a father the administrator of a child's estate when both parents were vying for the assignment. Rid of its context-specific details, this statute was a paradigmatic sex classification: males were treated differently from females.

Following Murray's road map, Ginsburg's principal arguments on behalf of Sally Reed were that sex classifications are discriminatory just like race classifications and so should also be subject to strict scrutiny, but even if the Court was unwilling to affirm this equation, automatically favoring fathers over mothers as estate administrators was "patently unreasonable." In other words, against the traditional grain that sex classifications are innocuous or even protective of women, Ginsburg argued that this one was bad discrimination. She didn't expect to convince the Court to use strict scrutiny the first time around, but she devoted the bulk of her brief—about fifty pages—to that argument, planting the seed for future consideration. She reserved six pages for the second argument, which was textbook reasonableness analysis. In between, she floated a one-paragraph pitch for "heightened" or *intermediate scrutiny*.

I teach torts—civil wrongs—to first-year (1L) law students. The reasonableness standard permeates the subject. Did he reasonably fear for his life? Was she reasonable to drive through the intersection as she did? Did they perform the surgery consistent with the standard of care, that is, of a reasonably prudent specialist? Was the product reasonably designed? My 1Ls would find every bit of Ginsburg's analysis familiar, including as it musters evidence from common sense, experience, data

177

and statistics, and a moral position against the state's claim that women are "mentally inferior" to men and that men have "greater 'business experience.'"

Writing for an unanimous, still-all-male court, Chief Justice Warren Burger found for Sally Reed on Ginsburg's reasonableness argument. The opinion recites the standard rational basis rule that classifications "must be reasonable, not arbitrary, and must rest upon some ground of difference having a fair and substantial relation to the object of the legislation, so that all persons similarly circumstanced shall be treated alike." Then, applying this rule, it concludes that the state's "arbitrary preference established in favor of males" is unconstitutional because women routinely serve as administrators, and no evidence was presented that they can't do the job as well as men. In other words, men and women are similarly situated as to the matter of handling decedents' estates, yet men were being privileged in the delegation of assignments.

Critical was the finding that administrative convenience alone would no longer suffice to justify sex discrimination. This meant that the state couldn't prefer fathers just because this might save judicial resources. The Court also decided that going forward, states could no longer rely on the courts' historical practice of coming up with plausible rationales for sex-based measures. Instead, the states themselves would have to provide the rationales. In all these ways, *Reed* broke the dam.

"A New Direction"

A year later, Ginsburg took the next step with another famous brief in the case *Frontiero v. Richardson*. *Frontiero*, with which I introduced this chapter, presented the question whether a military policy that automatically provided a housing allowance and medical benefits to the wife of a serviceman but not to the husband of a servicewoman violated the Equal Protection Clause. Sharron Frontiero and her husband, David

Cohen, were a role-busting couple. Both had jobs, but Sharron, who was in the Air Force, was the primary breadwinner. When she wasn't given the allowance and benefits for her spouse, as she put it later, "I walked into a lawyer's office and said, 'Help me get my money.'" Ginsburg wasn't that lawyer—Frontiero was represented by the Southern Poverty Law Center—but it was one of the cases Ginsburg selected to advance the cause as an *amicus curiae*, meaning as a friend of the court. Like her brief in *Reed*, this one isn't only a historical treasure; it also remains an invaluable template for practitioners.

Ginsburg again argued that sex classifications that unnecessarily curtail women's rights aren't innocuous exercises of the police power but rather discrimination in the bad sense of that word. She credited the Court's progressive decision in *Reed* as "marking a new direction" but emphasized that "the distance to equal opportunity for women in the United States remains considerable" and argued that *Reed* alone wasn't going to close the gap because state and lower federal courts weren't clear on what it required. As a result, she said, "Absent firm constitutional foundation for equal treatment of men and women by the law, women seeking to be judged on their individual merits will continue to encounter law-sanctioned obstacles." Citing Murray's analogy of race to sex, Ginsburg argued,

> Legislative discrimination grounded on sex, for purposes unrelated to any biological difference between the sexes, ranks with legislative discrimination based on race, another congenital, unalterable trait of birth, and merits no greater judicial deference. The time is now ripe for this Court to repudiate the premise that, with minimal justification, the legislature may draw a "sharp line between the sexes," just as this Court has repudiated once settled law that differential treatment of the races is constitutionally permissible. *Amicus* is asking this Court to add legislative distinctions based on sex to the category of "suspect" classifications.

Finally, as she had in *Reed*, Ginsburg argued that if the Court still wasn't ready to move to strict scrutiny, denying spousal benefits to female members of the military in any event was unreasonable; and she devoted another paragraph to the suggestion that the Court consider an "intermediate" or in-between standard for sex cases.

Oral argument before the Court is normally reserved for the parties, but the justices granted Ginsburg special leave to make the case orally that she had made in writing. It was the first of five successful arguments she made in that forum before she ascended to the federal bench.

Sharron Frontiero got her money—it was an eight-to-one decision in her favor—but there was a catch for Ginsburg: there was no majority behind a single opinion, meaning the justices didn't agree on the reason for their positions in Frontiero's favor. A plurality of four justices—William Brennan, William O. Douglas, Byron White, and Thurgood Marshall (who had been appointed to the Court in 1967)—borrowed liberally from Ginsburg's brief, finding that because "sex, like race and national origin, is an immutable characteristic determined solely by accident of birth," measures that differentiate on that basis to curtail women's civil rights should be subject to strict scrutiny. A different plurality of four—Potter Stewart, Lewis Powell, Warren Burger, and Harry Blackmun—found that dialing up the scrutiny was unnecessary because granting spousal benefits to servicemen but not to servicewomen was unreasonable under *Reed*.

Between 1973 and 1975, the Court took on more sex discrimination cases, which, per Ginsburg's own synonymous use of the words *sex* and *gender*, the justices were by then mostly calling *gender discrimination*. Some of these cases upheld and others rejected the sex-based classifications at issue. The justices continued to disagree about whether the appropriate review standard should be *Reed*-style reasonableness or strict scrutiny, with Brennan, Douglas, and Marshall dissenting in favor of the latter whenever a case was decided on the former. At the same time, some of the cases percolating in state and lower federal courts were being decided using Ginsburg's alternative standard of review—intermediate scrutiny.

In 1976, in *Craig v. Boren*, Brennan finally put together a five-vote majority for that alternative standard. At issue in *Craig* was the constitutionality of a state statute that changed the age of majority from eighteen for females and twenty-one for males to eighteen for both, but retained the old distinction for purposes of purchasing 3.2 percent "nonintoxicating" beer. In other words, women could buy this beer three years before men could. Writing for the majority, Brennan read *Reed* as already standing for heighted and intermediate scrutiny. This wasn't entirely right because it hadn't formally shifted the burden to the government to justify sex classifications, nor had it dialed up the requirements that far. But it was consistent with Ginsburg's longer-term goal.

Intermediate scrutiny requires the person complaining about an allegedly discriminatory law to prove that the measure distinguishes among people on the basis of sex and that they're a member of the outgroup. After they've done this, the burden shifts to the government to prove that the measure is motivated by an important—but not necessarily compelling—interest, and that the discrimination is substantially related to its achievement. It's like strict scrutiny in that the government, not the challenger, has the formal burden of justifying the classification. It's unlike strict scrutiny in that it's less likely to be fatal. Reflecting both the analogies and the distinctions, the idea is that sex classifications are more likely than race classifications to be based in real biological differences; in other words, they're less likely to be antagonistically motivated.

As applied to the facts of *Craig*, the majority agreed that traffic safety—the state's rationale for distinguishing males from females—was an important objective; but it rejected the government's argument that allowing females to buy beer at an earlier age was substantially related to its achievement. The majority read the evidence concerning consumption, driving, and accident patterns—which were then, and remain today, biased against adolescent and young adult males—to be insufficient to warrant the state's sex-based line-drawing. That the state classified the beer in question as "non-intoxicating" didn't help its cause.

William Rehnquist, a future chief justice, dissented from the majority's decision to adopt intermediate scrutiny for sex classifications on the grounds that the new standard was vague and lent itself to results-oriented decision-making:

> How is this Court to divine what objectives are important? How is it to determine whether a particular law is "substantially" related to the achievement of such objective, rather than related in some other way to its achievement? Both of the phrases are so diaphanous and elastic as to invite subjective judicial preferences or prejudices relating to particular types of legislation, masquerading as judgments whether such legislation is directed at "important" objectives or, whether the relationship to those objectives is "substantial" enough.

Because of or despite these issues, *Craig v. Boren* ended up being revolutionary in three ways that are relevant to our understanding of sex discrimination law today.

First, in formally shifting the burden to the government to justify sex classifications and heightening the standard it needed to meet, *Craig* effectively required policymakers to "choose either to realign their substantive laws in a gender-neutral fashion, or to adopt procedures for identifying those instances where the sex-centered generalization actually comported with fact." As a practical matter, the costs of defending a sex classification were such that few regulators would take on the burden—even when it could have been met. Litigation aversion arguably did more than anything else to advance the cause of sex equality in the "we are all the same" sense of that term. Even those who were indifferent to women's rights were incentivized to get with the program.

Second, *Craig* applied the heightened standard of review that had been developed to address the curtailment of women's rights also to address the curtailment of men's rights. Although this extension to men was originally justified by its utility for taking down the artificial barriers that burdened women—on the view that separate-spheres ideology created a complementary scheme—it would eventually come to be applied in a gender-neutral way, giving men a claim to equal treatment independent of concerns about women's rights.

Third, over the objection of the dissenters, the liberal majority treated the real sex differences in the case as "'archaic and overbroad' generalizations." Over time, this phrase would morph into "myths" and "stereotypes." In *Craig* itself, the evidence was about young men and alcohol: that they're more likely than females to drink and drive and that they're bigger risk takers. A year later, in *Dothard v. Rawlinson* (1977), it was about the relationship of height and weight to physical strength, and of sex to sexual attraction. At issue in *Dothard* was the constitutionality of a state's prisons policy that restricted jobs involving contact with male prisoners to individuals—whatever their sex—who were at least five feet, two inches tall and 120 pounds. The Court found that the policy discriminated against women because they're more likely than men to be shorter and lighter and the state didn't provide evidence of the relationship of those two traits to strength. Brennan concurred in part to say that the policy was based in "ancient canards about the proper role of women" and unestablished claims that "that they are physically less capable of protecting themselves and subduing unruly inmates." He also dismissed as "myth" the notion "that women, wittingly or not, are seductive sexual objects." (It's not a good idea to assume that judges will take even obvious facts as given, especially in this context.)

"An Exceedingly Persuasive Justification"

Two years later, in *Personnel Administrator of Massachusetts v. Feeney* (1979)—a case about the constitutionality of giving preferences to veterans in hiring—the Court paused to summarize the state of its views on equal protection doctrine. In the process, it characterized the three tests that apply to judging the constitutionality of measures that classify people according to their traits and experiences. That characterization has had a lasting impact on sex discrimination law.

The Court began with the basics, including the default rational basis test that applies, for example, to age classifications:

> The equal protection guarantee of the Fourteenth Amendment does not take from the States all power of classification. Most laws classify, and many affect certain groups unevenly, even though the law itself treats them no differently from all other members of the class described by the law. When the basic classification is rationally based, uneven effects upon particular groups within a class are ordinarily of no constitutional concern. The calculus of effects, the manner in which a particular law reverberates in a society, is a legislative, and not a judicial, responsibility. In assessing an equal protection challenge, a court is called upon only to measure the basic validity of the legislative classification.

It then noted the exceptions to this legislation-friendly default: Any racially discriminatory measure requires an "extraordinary justification." For sex discrimination, an "exceedingly persuasive justification" is needed. If you don't see the difference between an "extraordinary" and an "exceedingly persuasive" justification, you've spotted the problem or, depending on your perspective, the brilliance of this particular gloss.

Because *Feeney* itself didn't involve a sex classification, the "exceedingly persuasive justification" requirement wasn't formally attached to the intermediate scrutiny rule until three years later, in *Mississippi University for Women v. Hogan* (1982). At issue in *Hogan* was the constitutionality of Mississippi's policy setting aside one of its nursing schools for females only. A male applicant challenged the policy on the ground that it discriminated against him on the basis of his sex. I'm going to quote the relevant part of the majority's opinion in full because it shows how the Court used the requirement together with its characterization of real sex differences as "archaic and overbroad generalizations" to begin to dial up intermediate scrutiny.

We begin our analysis aided by several firmly established principles. Because the challenged policy expressly discriminates among applicants on the basis of gender, it is subject to scrutiny under the Equal Protection Clause of the Fourteenth Amendment. That this statutory policy discriminates against males, rather than against females, does not exempt it from scrutiny or reduce the standard of review. Our decisions also establish that the party seeking to uphold a statute that classifies individuals on the basis of their gender must carry the burden of showing an "exceedingly persuasive justification" for the classification. The burden is met only by showing at least that the classification serves "important governmental objectives and that the discriminatory means employed" are "substantially related to the achievement of those objectives."

Although the test for determining the validity of a gender-based classification is straightforward, it must be applied free of fixed notions concerning the roles and abilities of males and females. Care must be taken in ascertaining whether the statutory objective itself reflects archaic and stereotypic notions. Thus, if the statutory objective is to exclude or "protect" members of one gender because they

185

are presumed to suffer from an inherent handicap or to be innately inferior, the objective itself is illegitimate.

If the State's objective is legitimate and important, we next determine whether the requisite direct, substantial relationship between objective and means is present. The purpose of requiring that close relationship is to assure that the validity of a classification is determined through reasoned analysis rather than through the mechanical application of traditional, often inaccurate, assumptions about the proper roles of men and women.

In other words, anyone who maintains a sex- or gender-based policy should be prepared to defend it with an "exceedingly persuasive justification." A justification will not meet this standard unless it's established, using acceptable evidence—evidence that isn't based in "archaic and stereotypic notions" about the differences between males and females—that the policy was motivated "at least" by an "important governmental objective." The addition of the qualifier "at least" signals that only strict scrutiny's requirement of a "compelling" interest would guarantee safe harbor. The justification will separately fail if there's not a "direct," "close," and "substantial" relationship between that interest and the sex classification that is the subject of the challenge. When the Court applied this part of the test to the facts of the case, it added the qualifier "necessary"— again from strict scrutiny, concluding that "the record in this case is flatly inconsistent with the claim that excluding men from the School of Nursing is necessary to reach any of MUW's educational goals." The Court made clear that its intention was to force policymakers to engage in "reasoned analysis" *before* they decide to regulate on the basis of sex.

As applied, Mississippi's policy failed both prongs of the test. The Court found that the state's interest in a single-sex nursing school was illegitimate since female nurses predominated in the profession. It also found that excluding males wasn't substantially related to the program's educational goals because the state allowed men to audit classes at the school, and because the school used the same teaching styles in its single-sex and coed classrooms. In sum, the Court found that the state's objective was a pretext reflecting stereotypes about women as nurses.

The majority's opinion in *Hogan* was written by Associate Justice Sandra Day O'Connor, whom President Ronald Reagan had appointed to the Court the previous year. O'Connor grew up on a cattle ranch in Arizona, went to Stanford University at age sixteen to study economics, and then to Stanford Law before returning to Arizona, where she was an assistant attorney general, the first woman to serve as majority leader of the state senate, and a state court judge. She made an exceedingly persuasive case for being the first woman in history to be nominated and confirmed to the Supreme Court.

When O'Connor joined the Court's liberal wing in *Hogan*, it was as a conservative feminist. Having had many of Ginsburg's life experiences, she supported the ongoing equality effort, but she was skeptical of affirmative action. Affirmative action is positive discrimination, or, put differently, discrimination that's designed *to support*, not *to curtail*, a group's rights and opportunities. She imagined a day not so long into the future when the law could be race-blind. As she wrote in her opinion for the majority in *City of Richmond v. J. A. Croson Company* (1989), the case in which the Court decided that all race-based measures—whether they curtail or support rights—would be strictly scrutinized, "the 'ultimate goal'" is "eliminat[ing] entirely from governmental decisionmaking such irrelevant factors as a human being's race." She didn't hold the same view about sex.

United States v. Virginia (VMI)

Lawyers thrive on analogies and distinctions. I say: A is like B, not like C, so A should be treated like B. You respond: A is actually not like B, it's like C, so A should be treated like C. A specific example: I say sex is like age, not like race, and so sex should be treated like age. You respond, sex is actually not like age, it's like race, so sex should be treated like race.

In cases from *Craig* to *Hogan*, the Court made clear that it would embrace the sex-to-race analogy but that it would generally reject this standard opportunity for distinction. That is, I can argue that as black people's rights have been curtailed because of false assumptions about their inferiority, so too have women's rights; and I can produce admissible evidence—from history and, if necessary, also from biology and statistics—in support of that argument. But I'm on shaky ground if I argue that because of biological differences between males and females, sex is actually more like age than like race, because my evidence of this, my proof—*regardless of its quality and quantity*—is in danger of being described as myth and stereotype. Remember that as it was ridding the law of the artificial barriers that kept women down, the Court chose to describe as "archaic and overbroad generalization"—and so as legally irrelevant—the (true) facts that adolescent boys and young men drink more, drive more after they drink, and are bigger risk takers than girls; that height and weight have a bearing on strength; that females are sexually attractive to males (maybe especially to imprisoned men); and so on.

After *Hogan*, it seemed it would be just a matter of time before the Court would formally hold that sex is like race for all purposes (even though it's actually not), and that we would begin to use strict scrutiny to evaluate all sex classifications (as we do all race classifications). True to its design, this scrutiny would be fatal—none or almost none would survive—in part because we would continue to reject even compelling sex-differences evidence as legally irrelevant.

Only it didn't turn out that way. In between *Hogan* (1982) and *United States v. Virginia* (1996)—which is today the leading case on the question of the constitutionality of sex classifications—the Court's composition changed, and it decided a set of cases that would unsettle the justices' commitment to this trajectory. On composition, the Court lost Justices Burger, Marshall, and White and gained Justices Antonin Scalia (1986), Clarence Thomas (1991), and Ruth Bader Ginsburg (1993). On cases, I'll mention two, *J.E.B. v. Alabama* (1994) and *Adarand Constructors v. Peña* (1995), before turning to *United States v. Virginia*.

At issue in *J.E.B. v. Alabama* was the constitutionality of a prosecutor's use of sex-based peremptory challenges to exclude women from jury service. Peremptories are the limited number of free exclusions—those that normally don't need to be justified—the parties get during jury selection. The object is to allow the party exercising them to exclude a potential juror they suspect, but cannot prove, might be biased.

The majority in *J.E.B.*, which included O'Connor and Ginsburg, held that like race-based peremptories, which are illegal, sex-based peremptories violate the Equal Protection Clause. In *J.E.B.*, the prosecutor's challenges were based on the assumption that prospective female jurors might be biased against men in paternity and child support cases. Writing for the majority, Justice Harry Blackmun insisted that stereotypes about "gender simply may not serve as a proxy for bias." In a footnote he added that "gender stereotypes" are an "impermissible" basis for "gender classifications" because—even if they're accurate generalizations—they're "likely to stigmatize as well as to perpetuate historical patterns of discrimination."

Using the word *gender* as a synonym for *sex* as Blackmun had, O'Connor wrote separately to express her unease with this approach to evidence about real sex differences:

> We know that like race, gender matters. A plethora of studies make clear that in rape cases, for example, female jurors are somewhat more likely to vote to convict than male jurors. Moreover, though there have been no similarly definitive studies regarding, for example, sexual harassment, child custody, or spousal or child abuse, one

need not be a sexist to share the intuition that in certain cases a person's gender and resulting life experience will be relevant to his or her view of the case. "'Jurors are not expected to come into the jury box and leave behind all that their human experience has taught them.'" Individuals are not expected to ignore as jurors what they know as men—or women.

Today's decision severely limits a litigant's ability to act on this intuition, for the import of our holding is that any correlation between a juror's gender and attitudes is irrelevant as a matter of constitutional law. But to say that gender makes no difference as a matter of law is not to say that gender makes no difference as a matter of fact.

Nodding to the Court's race cases, O'Connor described the majority's position as a statement "about what this Nation stands for, rather than a statement of fact." In other words, the Court's characterization of even real sex differences as myth and stereotype was a policy choice to ignore the evidence; doing so was deemed necessary to achieve sex equality. She emphasized that this policy choice "is not costless"—*including for women*—and she worried about the consequences if sex was finally fully tracked to race:

> Will we, in the name of fighting gender discrimination, . . . preclude [a battered wife] from using her peremptory challenges to ensure that the jury of her peers contains as many women members as possible? I assume we will, but I hope we will not.

Scalia wrote a dissenting opinion that was joined by Rehnquist and Thomas. In a footnote responding to Blackmun, Scalia focused on the evolution of the terminology associated with *sex* and *gender* and marked how he would use the words:

Throughout this opinion, I shall refer to the issue as sex discrimination rather than (as the Court does) gender discrimination. The word "gender" has acquired the new and useful connotation of cultural and attitudinal characteristics (as opposed to physical characteristics) distinctive to the sexes. That is to say, gender is to sex as feminine is to female and masculine is to male.

He then elaborated on the point that worried O'Connor. Describing the majority's opinion as "an inspiring demonstration of how thoroughly up-to-date and right-thinking we Justices are in matters pertaining to the sexes (or as the Court would have it, the genders)," he declared that "unisex is unquestionably in fashion" and emphasized, as O'Connor had in her concurrence, the costs of that fashion:

> The Court's fervent defense of the proposition *il n'y a pas de différence entre les hommes et les femmes* (it stereotypes the opposite view as hateful "stereotyping") turns out to be . . . utterly irrelevant. Even if sex was a remarkably good predictor in certain cases, the Court would find its use in peremptories unconstitutional.

The following year, O'Connor wrote the majority opinion for the Court in *Adarand Constructors v. Peña* (1995), which affirmed that strict scrutiny attaches to all race classifications—including racial preferences—regardless of whether they're promulgated by the states or the federal government. The Court also decided that going forward, proof of a present or future injury would be necessary to sustain race-based affirmative action. The desire to remedy past discrimination wouldn't be enough. Given that race is the paradigm suspect class for purposes of strict scrutiny, *Adarand* signaled that this strictest approach to strict scrutiny was also on the horizon should sex eventually come under its auspices.

Whatever your position on race-based affirmative action, after

Adarand the race-sex analogy and the prospect of that long-awaited final move to strict scrutiny in sex cases had lost its luster.

The matter came to a head in *United States v. Virginia* (1996). The case involved a challenge to the constitutionality of the state's policy of admitting only males to the Virginia Military Institute (VMI). On behalf of the United States, the Clinton administration filed a brief arguing that it was time to "hold that 'strict scrutiny is the correct constitutional standard for evaluating classifications that deny opportunities to individuals based on their sex.'" This was the culmination of Justice Ginsburg's life's work on sex and gender discrimination, the opportunity she'd been seeking since she took the baton from Pauli Murray.

Ginsburg authored the majority opinion that garnered six votes, including O'Connor's, against VMI's exclusionary policy. Its principal passages are iconic in the history of sex in law for their articulation of what my Duke Law colleague Neil Siegel calls Ginsburg's "constitutional vision"—"equal citizenship stature." As Ginsburg herself put it, "equal" or "full" citizenship stature means "equal opportunity" "to aspire, achieve, participate in and contribute to society based on [a person's] individual talents and capacities." I'm sure I'm not the first to notice that her opinion is a complete response to Joseph Bradley's concurrence in *Bradwell v. Illinois*—refuting his constitutional vision as well as his treatment of the exceptional woman.

Ginsburg's *VMI* opinion also appears to be a response to how equal protection law had developed in the three short years from 1993, when she joined the Court, through 1996 when *VMI* was decided. In that period, the justices had been debating the costs and benefits of sex blindness—*il n'y a pas de différence entre les hommes et les femmes*—at the same time as they were debating the costs and benefits of race blindness, that is, of "eliminat[ing] entirely from governmental decisionmaking such irrelevant factors as a human being's race." Given the opportunity finally to cement the analogy of sex to race and formally to take sex over the line to strict scrutiny, Ginsburg chose instead to distinguish between the two. In so doing, her opinion not only froze in place O'Connor's approach from *Hogan*, but it added a list of objectives that could make it possible for a sex-based classification to pass muster—objectives that, after *Adarand*, would be insufficient as justifications for race-based classifications.

She began by reaffirming the applicability of intermediate scrutiny to the evaluation of sex classifications:

> Without equating gender classifications, for all purposes, to classifications based on race or national origin, the Court, in post-*Reed* decisions, has carefully inspected official action that closes a door or denies opportunity to women (or to men). To summarize the Court's current directions for cases of official classification based on gender: Focusing on the differential treatment or denial of opportunity for which relief is sought, the reviewing court must determine whether the proffered justification is "exceedingly persuasive." The burden of justification is demanding and it rests entirely on the State. The State must show "at least that the [challenged] classification serves 'important governmental objectives and that the discriminatory means employed' are 'substantially related to the achievement of those objectives.'" The justification must be genuine, not hypothesized or invented *post hoc* in response to litigation. And it must not rely on overbroad generalizations about the different talents, capacities, or preferences of males and females.

Then she emphasized the distinction between race and sex classifications, using language that would become ironic in and beyond the law:

> The heightened review standard our precedent establishes does not make sex a proscribed classification. Supposed "inherent differences" are no longer accepted as a ground for race or national

origin classifications. Physical differences between men and women, however, are enduring: "[T]he two sexes are not fungible; a community made up exclusively of one [sex] is different from a community composed of both."

"Inherent differences" between men and women, we have come to appreciate, remain cause for celebration, but not for denigration of the members of either sex or for artificial constraints on an individual's opportunity. Sex classifications may be used to compensate women "for particular economic disabilities [they have] suffered," to "promot[e] equal employment opportunity," to advance full development of the talent and capacities of our Nation's people. But such classifications may not be used, as they once were, to create or perpetuate the legal, social, and economic inferiority of women.

This retrenchment meant that the door remained open for a sex classification to survive an equal protection challenge, but not for VMI's admissions policy. As to that, the Court found the "Commonwealth has shown no 'exceedingly persuasive justification' for withholding from women qualified for the experience premiere training of the kind VMI affords." The majority's repeated use of the term "exceedingly persuasive justification," together with its rejection of the state's sex-differences evidence as "overbroad generalizations," prompted a dissenting Scalia to charge that the Court had replaced intermediate scrutiny with this gloss, which—notwithstanding Ginsburg's now-iconic text—he read as "indistinguishable from strict scrutiny." Ted Olson, who argued the case for Virginia, told me that he remembers being frustrated at oral argument that he wasn't able to get any traction with the justices with his evidence in support of single-sex educational options, in the form of expert testimony about "gender-based developmental differences," "differences between men and women in learning and developmental needs," and "psychological and sociological differences."

Scalia and Olson were right: it's difficult to square the policy-driven rejection of real sex-differences evidence as "overbroad generalizations" with the promise that scrutiny of sex classifications won't always be fatal. Scalia's dissent is valuable for the insights it offers into the conversations among the justices, perhaps especially as the famous opera buddies exchanged misgivings and honed their respective positions. Harking back to O'Connor's concurrence in *J.E.B.*, and perhaps also to the incongruity of seeing the *distinction* between race and sex included in an opinion authored by Ginsburg, who as an advocate had built her legacy on the *analogy*, he wrote of "the Court's unease" with the "consequences" of the decision "despite its unwillingness to acknowledge them."

Ginsburg, O'Connor, and their colleagues in the majority were right too: unlike "[s]upposed 'inherent differences'" among the races, "[p]hysical differences between men and women" are "enduring: '[T]he two sexes are not fungible; a community made up exclusively of one [sex] is different from a community composed of both.'" Physical differences between men and women aren't just enduring; they're also "cause for celebration." Beyond the basics of human reproduction, communities "composed of both" are the better for being so. Finally, the law shouldn't "make sex a proscribed classification" for these timeless reasons and because regulating on the basis of sex may be necessary "to compensate women 'for particular economic disabilities [they have] suffered,' to 'promot[e] equal employment opportunity,' [and] to advance full development of the talent and capacities of our Nation's people." In other words, rejecting sex blindness in law—and allowing policymakers to focus on evidence-based norms, not just exceptions—is essential to being able to deal with the concerns of the real world.

Indeed, as we encounter the many modern situations in which sex matters—because it's good, because (like age) it just is, and because (like race) it's still a problem—having the option of sex-based tools to address sex-based differences will be invaluable. As we saw in chapter 5, we're in a position today to face our policy challenges—and our routine social interactions—with much better evidence about sex differences than we had in the past. We have the potential to make life better for all human beings, including for people in the LGBTQ communities who exist in sexed bodies like the rest of us, but only if we can use this evidence. And so it makes sense to ask again, as O'Connor did in *J.E.B.* and Scalia did in *VMI*, whether the costs of treating sex like race—as an irrelevant factor in decision-making—are worth bearing. I say no.

Ruth Bader Ginsburg did incalculable good. It's a testament to her life's work, and to the strategic decisions she made along the way, that in wide swaths of the world women have and control their own money and have independent lives in the public sphere. Men take parental leave after the birth of a child. Gay people are free not only to act on their sexual orientation but also to get married. And transgender people are—despite the current backlash—well on their way to full citizenship stature. Ginsburg's contributions were about removing the "artificial barriers"—meaning gender in its socially constructed sense—from the law of sex so that everyone could get to that place. Her work was central to the fact that when we think *sex* today, the second thing that comes to mind is equality, and that our new challenge, which will be addressed with the tools she honed, is sex itself.

As we engage with this new challenge, it serves us well to return to Pauli Murray. In 1971, as Ginsburg was beginning her work with the ACLU, Murray, who was twice denied admission to the University of North Carolina at Chapel Hill because of her race (in 1938 and 1951), became a tenured member of the faculty at Brandeis University, where she cofounded the American Studies Department. When her beloved partner, Irene Barlow, died of breast cancer two years later, Murray entered the seminary. In 1977 she became the first black woman—and one of the first women—to be ordained as a priest in the Episcopalian Church. She celebrated her first Eucharist in Chapel Hill, reading "from the Bible that had belonged to her grandmother (Cornelia Smith), from a lectern that had been given in memory of the woman who owned Cornelia (Mary Riffin Smith)."

For my law school colleagues Jeff and Sarah Powell, who are Episcopalian, Murray has long been a sung hero. As Jeff put it to me, "She was a person who knew suffering, personally and as a Black American and a woman, and yet she was full of love and saw Christ's work as the reconciliation of all human beings to one another and to God." He added that the date of Murray's death, July 1, is regularly celebrated by their church as Catholics celebrate saints: "She is 'on the calendar.'"

Since Murray's death in 1985, she's also become a sung hero to the LGBTQ communities as her archives, in her own words, show not only that she was gay, but that she also struggled, from childhood, to reconcile her female sex with her inner sense of herself as male.

Murray's life and work tell us how important it is to see people as humans first and last, but in between not to miss the ways in which their different circumstances—including their race, their sex, their sexual orientation, and their gender identity—affect their opportunities to live fulfilled lives. She was a person who knew suffering on all of these accounts and more. But she also knew love, intelligence, and grace (both kinds) and so could see others clearly. Equality was central to her sense of the opportunity to live a fulfilled life, including as it can facilitate freedom and necessaries, but only if it exists beyond theory and makes life better for regular people on the ground they actually inhabit.

8

The Politics of Sex and Gender

On MAY 17, 2019, the Democrats in the House of Representatives brought the Equality Act to a vote on the floor of the chamber. As we saw in the introduction and then in more detail in chapter 3, the act redefines *sex* in federal law to be sexual orientation, gender identity, sex stereotype, and sex characteristics, but apparently not biological sex itself. Indeed, its proponents are adamant that we not say "biological sex." Beyond that, it takes the step the Supreme Court in *United States v. Virginia* wouldn't: It requires sex blindness in all federally funded programming as well as in all public accommodations. It defines public accommodations broadly to include all goods and services whether these are provided in physical spaces or not. Directly or indirectly, its passage would close the door to any exceptions that would allow us to continue to celebrate sex, to compensate women for historical discrimination, and—as Justice Ginsburg would have allowed—"to advance full development of the talents and capacities of our Nation's people."

The Equality Act looks a lot like a Christmas tree with a present under it for everyone in the progressive family. Academics who'd been digging in on how the male-female binary is all a social construction got what they were hoping for. Advocates who hadn't managed to get the Supreme Court to hold that sex is all myth and stereotype got theirs too. Gay and trans people got a lot of stuff they legitimately need—especially assurances that in the United States no one can discriminate against them as they go about the business of living their lives. But all of this is packaged—in the preamble—in a claim about how women need the legislation too, because we aren't treated equally when we buy things in stores and restaurants and navigate entertainment and transportation.

As I was reading it for the first time, all I could think of was that accessing public accommodations isn't a problem for women given different federal laws, protections at the state level, egalitarian norms, and market forces. It seemed that the Equality Act wasn't being pitched honestly, as legislation to help gay and trans people. Rather, its backers seemed to be spinning it as pro-woman to get our support, or at least not to arouse our animosity, and to set us up as the engine that could bring everyone over the hill to equal citizenship status.

As we saw in chapter 7, civil rights movements work like a relay—and also like a ladder where you bring other people up behind you: black people and people of Japanese descent to women to gay people to trans people. As such the strategy behind the Equality Act was both obvious and good. It was true then as it is now that gay and trans people need the protection of federal law given the many states that still allow discrimination against them because of their sexual orientation or gender identity. Good on the Dems for doing this work.

But it was in no way true that the bill as written was good for women. Women are *not* construction sets. Whatever our gender identity, our distinct physical, cultural, and interpersonal experiences are not myth and stereotype. It's important for our health, our safety, and our welfare—and yes, also for our equality and our liberty—that the door that Justices Ginsburg and O'Connor and their colleagues on the Court left ajar for sex-smart lawmaking remains open. Let me repeat: getting gay and trans people to full citizenship status is obvious and good, and women have always been the engine that could. That's a piece of our historical role. But using us and then undermining our interests is more than wrong. That self-proclaimed "experts in women" think differently doesn't change this for those of us who don't buy into their utopian project, and neither does their incantation of the movement mantra "What's good for trans women is good for all women." Saying it over and over again doesn't make it so.

Capitalizing on this familial fracture, Republican legislators introduced an amendment on the floor of the House that would have created an exception for girls' and women's sports. In effect, the proposed amendment was for a "yay" or "nay" on continuing the fifty-year-old legal commitment we know colloquially as "Title IX"—a regulation that, as interpreted by the courts, requires separate-sex teams as the way to ensure equality for female student-athletes. This was savvy gamesmanship since support for girls' and women's sports has become a proxy for support for female empowerment. The amendment failed along party lines: the Republicans voted for it, the Democrats against. H.R. 5 itself passed, also along party lines. Given that the GOP wouldn't have voted for the act anyway and that it had no chance in the Senate, it was all perverse political theater of the highest order.

Between these floor votes and the earlier Equality Act hearing, I coauthored an op-ed in the *Washington Post* with Martina Navratilova and Sanya Richards-Ross called "Pass the Equality Act, but Don't Abandon Title IX." We ended with a plea for common sense:

Sport is an unusual if not unique institution. It is a public space where the relevance of sex is undeniable, and where pretending that it is irrelevant, as the Equality Act suggests, will cause the very harm Title IX was enacted to address.

We support transgender women and girls and their right to equality, and we recognize their personal struggle. We don't worry that boys and men will feign transgender identity to gain an advantage. But we do hope that lawmakers won't make the unnecessary and ironic mistake of sacrificing the enormously valuable good that is female sports in their effort to secure the rights of transgender women and girls.

I explained my own debt to Title IX in an essay for Quillette that same spring, "A Victory for Female Athletes Everywhere." Here is a small excerpt.

My own story is a testament to the power of mandatory set-asides for female sport and to the value they create for girls and women that would not exist otherwise. I was the first female recipient of a full track scholarship to Villanova University in 1978, six years after Title IX passed into law. I was recruited because I was one of the best under-18 (U18) female half-milers coming out of U.S. high schools that year. Because even mediocre boys could and did run faster, had Title IX not forced colleges to create programs and set-aside funds for girls, I wouldn't have gotten that scholarship. And because my family was poor, I might never have gone to college. My life story would have been altered in innumerable ways. Most importantly for present purposes, I would not exist as someone with the leadership skills and experience to advocate either for clean sport or for equality for females. I would not exist as someone who could give back, certainly not in the way that I do as a law teacher, and certainly not in this global context. Title IX powered this outcome.

Importantly, it did this not only by affording me that first scholarship, but also by providing me with the same chance as the best boys coming out of high school at securing the longer-term benefits of participation in elite sport. Winning gave me confidence, including on a stage. Training for long-term goals taught me time management, independence, and goal orientation. Losing made me resilient. And traveling made me tough and sophisticated about the world, including about how to make my voice heard in traditionally male spaces. The same is true for many other girls and women for whom elite sport has also been something of an equalizer in a world that has long privileged boys and men.

Title IX also powered Sanya Richards-Ross. For about a decade in the early 2000s, she was the fastest female on the planet over 400 meters—one lap around the track—with an extraordinary ten global golds to her name. As of this writing, her 2006 American record still stands at 48.70 seconds. Born in Jamaica, Sanya became an American citizen as a child and benefited throughout her life in school—finally as a proud Texas Longhorn—from Title IX.

Martina made it without Title IX, but she wouldn't have gotten out of Czechoslovakia, and we wouldn't know her name, at least not as an elite athlete, if competitive tennis wasn't categorized by sex. Martina also has a lifetime of involvement with LGBTQ rights, lending her ideas, her voice, her prestige, and her money to related causes. Among other projects, she cofounded Athlete Ally, an organization that's active in this period in support of trans athletes.

In other words, three on-side women with relevant knowledge and experience announced their support for legislation to ensure comprehensive civil rights protections for gay and trans people. We flagged the fact that *the way these were being provided in this draft legislation* would undo existing civil rights protections for females. And we asked legislators to amend the draft to ensure that we didn't become collateral damage.

In response, trans advocacy organizations went into overdrive. They were joined by the traditional women's groups, including the National Women's Law Center, which by then had adopted the trans groups' position that "women" isn't about (biological) females or even (biological) sex. In those weeks around the Equality Act hearing and floor vote, the groups engaged in a coordinated public relations campaign that pushed out the movement mantras: Sex differences are "myth and stereotype." "Trans women are women, period" (often abbreviated as "TWAW, period"). "What's good for trans women is good for all women." Consistent with the redefinition of *sex* in the Equality Act, the "women" in their organizational names had become an umbrella concept that includes both males and females. If you resist this point publicly, as the ACLU's deputy director for transgender justice Chase Strangio put it to Martina and me later, via thinly veiled language in an essay in Medium, you're killing trans kids.

In May 2019, in the middle of this fraught period, I received an email from Neil Munro, an editor at Breitbart News. I have to pause here to say how disconcerting this was. I'm a lifelong Democrat, and Breitbart was Steve Bannon country in the heart of the Trump administration. I'm not on social media because of people like Strangio, some of whose words have a tendency to incite their own violence. If you want to reach me, my inbox is it. The notion that it would come to include a message from anyone at Breitbart goes a long way to describing the bizarre nature of the moment we're living in. Still, its substance was illuminating. Munro was looking for me to comment on what he saw as the *Post*'s failure to promote our op-ed—I declined. But he said of transgender people: "Of course they exist, albeit in very small numbers, and the vast majority of Americans are willing to live and let live. See the polling data I provide." He added that "what gives this tiny group such ambition (You must believe I am a woman!) and influence is the confluence of their diverse interests with that of other groups." For example,

the ACLU & progressive law professors like yourself have a huge professional & status incentive to join the progressive movement to liberate people from testosterone policing. Would your colleagues support you under those circumstances? You've seen the crap that Navratilova faces for stating the obvious.

Reflecting on this forced allyship problem, he added that the "direct-to-consumer power" the Internet affords outlets like Breitbart "does not exist for most professionals," including law professors and journalists in the legacy press "who rely on each other for opportunity, income & status." Then the point of grand strategy: "I suspect the Koch network would be happy for Trump to use this issue to win the 2020 election. The Koch donors really, really, really do not want him winning on labor supply and wages."

If you don't follow sex in sports because you don't care about sports, or women's sports, or whether some trans women and nonbinary males win some championships set aside for females, know that most advocates and politicians who are working the issue don't really care either.

The right fought equity in athletics from the beginning. It's still not fully committed, something the left enjoys pointing out whenever the occasion arises. That it was the GOP and not the Dems that invited me to Capitol Hill to testify in support of Title IX, and that it introduced the amendment to the Equality Act to save women's sports, was off-the-charts ironic. That they've since ignored my offers to assist in the development of a viable measure to these ends isn't surprising. But it will eventually make sense for them to bite the bullet, if only because there's a legion of former elite athletes with cultural and political power who pull red in the voting booth and are now demanding it. This includes Olympic gold medalist and sports broadcaster Donna de Varona and young Riley Gaines, the elite swimmer turned political activist from the University of Kentucky who was made to stand down for Lia Thomas at the 2022 NCAA National Championships.

The left can be equally disingenuous. True to the policy choice to ignore even real sex-differences, their leading voices repeatedly demonstrate disregard for competition, the biology of human performance, and the sex-based rationale for the female category. This rationale is obviously undermined by the demand that we ignore sex in the eligibility standards. For those of us who do care about sports, their disregard is hard to stomach, as there is no other plausible basis for the category—it's certainly not about affirming femininity, which would actually be myth and stereotype. As they erode the foundation, they say that it's okay if females don't win because what matters is participation, not the podium. It's been several years since I first heard this argument and my reaction is still, Wow! They really don't see us at all.

The fact is that people in both camps—those who claim to be "saving women's sports" and those who claim that real sex differences are "myth and stereotype"—are using the institution as a proxy for the ultimate questions: Whether sex still means sex. Whether we can still see it and say that it matters. Whether—as humans have done since time immemorial—we can continue to use it to sort people to meet individual and collective needs. If not in arenas as purely physical as athletics, then nowhere. *This* was the reason for the political theater in the nation's capital.

The Religious Right and the Radical Left

How did we get to this place where sex and gender are at the heart of our political life, infecting the media and big institutions, driving our political divisions to new heights, including to the point where they can affect elections? The story I've just described is certainly part of it. Despite the persistent investment gap between men's and women's sports, Americans in general, and on a bipartisan basis, care a lot about Title IX. But it's also about overreach by the left and the right and about the radical left's decision to go after the free exchange of ideas.

Before I get to these two points it's important to know the principal players and their priors. Their underlying commitments go far to explaining their positions and their uncompromising strategies.

In the United States, the far right is primarily the religious right. These believers start from the male-female binary: Life and protectible personhood begin at conception when (male) sperm meets (female) egg. We develop in utero and are born male or female. Whatever gender identity is, no one is born in the wrong body. To paraphrase Caster Semenya, everyone is as God made them and that is natural and good. Heterosexuality and heterosexual marriage complement the male-female binary and are also prescribed by nature and God.

For the religious right, gender role and gender expression should match sex as ways to keep everyone on the cultural script that prevails in the time and place. This script doesn't just exist for the individual so that they know their place, but also for the rest of us so that we know who we're dealing with. The gender script isn't stagnant, although the religious right is conservative in the literal "conserving" sense of that word. They may not be into nail polish on men but they're happy that Amy Coney Barrett is on the Supreme Court—something Joseph Bradley would never have countenanced. Barrett is an archetype of the modern, intelligent, conservative woman. She's secure in her faith, empowered by her physical femininity and by maternity and mother-hood.

The far or radical left is mostly composed of academics and trans rights advocates. I'm again using the word *radical* here in its literal sense—per Oxford Languages via Google—as "affecting the fundamental nature of something" or "advocating . . . thorough or complete political or social change." As we saw in chapter 3 on the answer to the question "What is sex?" from progressive advocacy, the radical left starts from the premise that sex isn't real, that the male-female binary is a social con-struction built from our sex characteristics which—depending on who you're talking to—are always or mostly on a spectrum, and which can be sorted however we'd like if we have unfettered bodily autonomy and access to medicine. (We *are* construction sets, in other words.) They add gender identity to the set of sex characteristics and identify it as most important of all.

From this stance, some of us *are* born in the wrong body—or a mis-matched one or just not the preferred one—since being in the right one isn't about reproduction or reflecting or respecting nature and God, but rather about embodying one's authentic self. Regardless, any focus on sex is inherently illegitimate. My pronouns and how I'm described on my child's birth certificate or on my passport are all about my inner sense of myself and how I want to be seen, not about anything others have the right to know. Sometimes all it takes to express authentic gen-der is self-declaration, or it may be a haircut and clothing, but those who need or want to use medicine—physicians, drugs, surgery—to embody their truth are entitled to do so. This entitlement is said to derive from human rights, civil rights, and individual liberties. Gender-affirming care is packaged together with abortion as an application of unfettered bodily autonomy.

Neither the religious right nor the radical left is close to repre-senting a majority of the people, so normally their respective views, each of which contains kernels of truth but are outliers as integrated philosophies, wouldn't be of concern. Normally we'd be free to engage with them or not as we wished. However, because of their cultural capital, both came to have political power, and so their views came to matter a lot.

Overreaching in Moments of Power

Gay people and transgender people have always existed. The word *transgender* may be new but the experiences it captures—including gender dysphoria, nonbinary identities, and cross-dressing—are there in the historical record. As I've maintained throughout this book, the question is not whether this is fact but rather how societies choose to treat gender diversity. A second fact is that societies, including ours, traditionally haven't treated it well. In the process, gay people and trans people have been denied their lives, their liberty, and their opportunity to pursue happiness—by which I mean the opportunity to live a fulfilled life.

In the name of the society's health, safety, and morals, our norms and laws historically criminalized not their existence as humans but their gender expression as gay or trans. You might see this described as "erasing" their transness or their sexuality. The Smithsonian Institution, which has compiled the history in the United States, notes that in the twentieth century, in addition to criminal sanctions to prevent "deviancy" on the grounds of nonconforming sexual orientation and gender expression, you could be subject to "electroconvulsive shock, lobotomy, drugs, and [forced] psychoanalysis." These barbaric tactics are now illegal, but persistent social and religious norms in many parts of the country continue to make it difficult for gay and trans people to be "out" as such. Some of the same people who believe that the *David* is pornographic, for example, also believe that transness is a mental illness that should be cured, and that none of this should be presented to children as a way to be.

In the last twenty years, gay and trans rights advocates have done extraordinary work for their respective causes. Here are just some of the watershed moments: In 2003, in a six-to-three decision in *Lawrence v. Texas*, the Supreme Court ruled that it's a violation of the Due Process Clause of the Fourteenth Amendment—its *liberty* guarantee again—to criminalize sexual relations among consenting adults. In 2011, gay people were permitted to serve openly in the military. In 2013, California and New Jersey, followed by other states, banned "conversion therapy"—the umbrella term for that set of despicable clinical practices that had historically been employed to "cure" gay and trans people of their "deviancy." Special attention was paid in those reforms to the misuse of conversion therapy on children by people—mostly clerics and doctors—acting for parents. In 2015, in a five-to-four decision in *Obergefell v. Hodges*, the Court ruled that it's a violation of the Due Process and Equal Protection Clauses to deny same-sex couples a marriage license. In 2020, in a six-to-three decision in *Bostock v. Clayton County*, the Court held that an employer violates the Civil Rights Act of 1964 when it treats an employee differently because of their sexual orientation or their gender identity. In 2021, trans people were permitted to serve openly in the military.

Credit for these developments goes to the advocacy groups, to the Democratic Party, and ultimately to centrists in both parties who get that it's irrational and cruel to pathologize homosexuality and gender diversity. If not before, certainly by the first two decades of the twenty-first century, all straight people knew gay people as members of their families, their communities, or their workplaces, and it was unbelievable that our predecessors could have criminalized and "treated" them as they had done. If not before, then by the second decade of the twenty-first century, a growing majority knew trans people, or at least of trans people, in the same way. As Michelle Goldberg of the *New York Times* put it in a podcast interview with Ezra Klein in 2022, "Everybody I know kind of knows people who have kids who are either transitioning or nonbinary, and maybe they're confused by that. But they're not hostile to it." Just as TV shows like *Modern Family* normalized gay people and gay families, shows like *Pose* opened our eyes to trans people's humanity and also to their pain. Blanca Rodriguez, Angel Evangelista, and Elektra Evangelista as characters—and the extraordinary actors who played them—showed us that this pain is exponentially worse when we reject them.

In the aftermath of *Obergefell* (2015), the left turned its attentions immediately to the matter of trans rights. The first version of the Equality Act appeared just a short month after that decision was handed down. This made political sense for those for whom this was the next commitment, and it was smart to take advantage of the momentum. And there was a lot of that. The Court's decision to legalize same-sex marriage may have come on a slim five-to-four majority, but Justice Anthony Kennedy's closing words ensured that "Love is Love" was that summer's refrain:

> No union is more profound than marriage, for it embodies the highest ideals of love, fidelity, devotion, sacrifice, and family. In forming a marital union, two people become something greater than once they were. As some of the petitioners in these cases demonstrate, marriage embodies a love that may endure even past death. It would misunderstand these men and women to say they disrespect the idea of marriage. Their plea is that they do respect it, respect it so deeply that they seek to find its fulfillment for themselves. Their hope is not to be condemned to live in loneliness, excluded from one of civilization's oldest institutions. They ask for equal dignity in the eyes of the law. The Constitution grants them that right.

Like many others that year and in the years that followed, I had the honor of standing up for two of my friends—for me it was Bill and Jonathan—and from the pulpit of a chapel reading these incalculably healing and hopeful words.

Consistent with his own civil rights priors, President Barack Obama's celebratory remarks on the lawn of the White House suggested that the Court's expansion of the scope of our constitutional liberty—its inclusion of more people within its terms—was good for the country: "[T]his ruling is a victory for America," he said. It "affirms what millions of Americans already believe in their hearts: when all Americans are treated as equals we are all more free."

But then the activists overreached. They used their influence in the Obama administration, on the Hill, and eventually in the Biden administration not just to add trans rights to rights for gay people and women and black people, but also to insist that the government say, consistent with their broader, underlying commitments: (1) Sex is or includes gender identity. (2) A transgender person is the sex they say they are whether they've transitioned physically or not. (3) Equity requires prioritizing trans people.

One of the most remarkable conversations I was part of in this period was with a member of the Biden administration. It fully captured the merging of sex and gender, the erasure of biology, the resumed subordination of females, and the convenient use of stereotypes. Asked about the administration's approach to the question "Who gets the last spot on the girls' team bus in the event there are two athletes remaining, a cisgender girl and a transgender girl?" the official's response, without hesitation, was "The trans girl, because she needs our support more."

The activists also targeted the private sector:

This is from the ACLU's Q&As on California law: "Q: I am nonbinary, and go to a gym or spa that segregates facilities by gender. Can the business choose which facilities I use? A: A gym, spa, or other

211

similar business can legally divide its locker rooms or other facilities into 'men's' and 'women's.' However, if your gender identity is non-binary, you can use the facility in which you feel most comfortable—the business doesn't get to choose."

This is from an article in the *New York Times* in 2019: "In a nod to transgender and nonbinary customers, Procter & Gamble said this week that it was removing the Venus symbol [♀], which has histori-cally been associated with womanhood and the female sex, from the wrappers of Always brand sanitary pads."

Companies now strongly encourage their employees to add pro-nouns to their signature lines—that is, to identify and announce themselves by their gender identity. For example, "Ashlee Brown, Pronouns: she/her/hers." The pitch is that doing this signals soli-darity with trans people, but it's also designed to have us signing on to the broader agenda. Some companies facilitated the process of conforming signature lines by providing their employees with a standard template that embedded a hyperlink (attached to the word *Pronouns*) to the Human Rights Campaign's (HRC) website. Whenever an employee sent an email, whether they knew it or not, they were sharing a link allying themselves to that advocacy group.

News organizations were lobbied to follow suit, and so you got disorienting articles with headlines like this one in the *New York Times* in 2022: "She Killed Two Women. At 83, She Is Charged with Dismembering a Third. Harvey Marcelin was charged with murder after a head was found in her Brooklyn apartment. Officials said it belonged to a dismembered body discovered in a shopping cart." I guarantee that most women who read the *she/her* pronouns in that headline felt a lot of cognitive dissonance until they got to "Harvey" and then had their intuition confirmed when they got to the part beyond the headline about how Harvey is male.

However you feel about these particular examples—at best I feel marginalized—you have to recognize that in a microsecond, historically speaking, sex and gender exploded onto the scene and the change was radical; it "affect[ed] the fundamental nature of" the first human and political taxonomy: Identifying people by their sex may be natural and adaptive but it's no longer courteous form. Menstruation is not about being female. Persons of the male sex are welcome in women's intimate spaces, even if they present as male, so long as their gender is nonbinary.

How did the radical left manage all of this, especially so quickly? Part of the answer is that its goals overlap with those of other social justice–oriented groups. Equality is undoubtedly a shared objective. To the extent their interests didn't overlap—because the devil is always in the details—allyship norms mostly prevailed: Like NATO, the idea is that they're there for each other. If someone doesn't play ball, they're out socially and professionally.

There were other factors too, including passion, excellent lawyers, money—seed money followed by corporate ESG/DEI money in exchange for good-company stickers from the HRC—media manipulation, progressive Twitter, and young people: future employees and voters, for whom that microsecond in historical time was their coming of age.

It may be personal to Joe Biden somehow, but his 2020 campaign declaration that "trans rights are the civil rights issue our time" and the White House's subsequent embrace of Dylan Mulvaney—then a twenty-something trans woman who, as a TikTok star before she had genital surgery, announced that she wanted to be able to wear hot pants in public so we all need to "normalize the bulge!"—suggest that his team was either tone deaf or making the strategic choice to dismiss the Bud Light crowd. As I write, I don't know if their calculation will pan out as they hope or if—like Bud Light's owner Anheuser-Busch, which also embraced Mulvaney—they'll pay for this choice. But I do know that much of this was bizarre to a lot of people on both left and right.

It's useful periodically to return to home base: whatever our gender (identity)—which runs the gamut from male to female to both to none—we all have one or the other (biological) sex. With the truly minuscule exception of people with ovotesticular disorder, even "intersex" people are male or female. Most of us don't have a problem with sex and a whole lot of us embrace it. *This is key: sex isn't only about discrimination and equality.* For a lot of us, it's also or even mostly about procreation, sexuality, beauty, our natural development and experiences over our lifetimes, and our morbidity and mortality patterns. We want to support people who are transgender, but "freeing" ourselves from sex is neither realistic nor attractive.

The radical left tries to marginalize this worldview—usually by dismissing it as "right-wing" or as "anti-trans hate"—the term that now seems to be replacing *transphobia* in the advocacy groups' lexicon; but that's wrong. *Knowing that sex is real and good isn't a partisan proposition. It's a human one.*

I'm Exhibit A in this book, but I'm in ample company. My liberal female friends from abroad who didn't grow up within the Anglo-American civil rights frame want to be free from sex discrimination, but the notion that sex itself is something we need to be liberated from is foreign to them. They're not for throwing the baby out with the bathwater. Neither are a lot of homegrown progressives. The comments sections associated with related stories and opinion pieces in the "mainstream" or "liberal" media—including in the *New York Times* and the *Washington Post*—are consistently full of such voices, including some who make a point of saying they're gay or trans, who reject the premise that you're anti-trans if you see and find value in sex, or that being a good ally to trans people requires sex blindness.

Michelle Goldberg of the *New York Times* is a prominent example.

She's written in support of trans people and trans kids, but about the backlash against radical feminism she said, "[T]he idea of making the gender binary, kind of irrelevant as a social organizing principle, I mean, that's pretty utopian. We've never seen anything like that in, as far as I can tell, the history of the world." And "every society is organized in some sense around the gender binary."

Andrew Sullivan has made a point of chronicling this radical over-reach. In a 2023 essay titled "Gay Rights and the Limits of Liberalism," he describes a speech he gave to the HRC when he was "a marriage equality pioneer." He remembers saying, "The goals of any civil rights movement should be to shut itself down one day. And once we get marriage equality and military service, those of us in the gay rights movement should throw a party, end the movement, and get on with our lives." This was right, he says, because liberalism "does not seek to impose meaning on everyone; it creates the space for individuals to choose that for themselves."

He then added:

> But in the movement I was once a part of, many, of course, were not liberals, let alone liberal conservatives—but radicals, who reluctantly went along with marriage equality, but itched to transform society far more comprehensively. And these radicals now control every-thing in the hollowed-out gay rights apparatus. Their main ticket item is a law that would replace biological sex with gender in the law, and remove protections for religious liberty: smashing the liberal settlement. Combine that with acute polarization in the Trump era, and information silos, so that many gays get their sense of reality from MSNBC and Elton John, and you can see how the spiral into illiberal madness began.

He concluded with the specifics of the disconnect between what he sees as the original bargain and the movement's current claims:

When majorities supported gay couples getting married, they did not thereby support having their daughters forced to shower next to biological males in locker rooms, or compete with them in competitive sports; they did not support teaching kindergartners that their bodies have nothing to do with whether they are boys or girls; they did not support using unapproved drugs on troubled children to arrest their puberty, and sterilize them for life; and they did not support schools transitioning their children into the opposite gender without their knowledge.

They didn't support these things because they have absolutely nothing to do with gay rights. And they didn't support them because trying to abolish sex differences in society—differences that are among the most well-established facts of human existence—is *insane*. Sure, many Americans were and are open to helping transgender kids be accepted, to treating trans adults with sympathy and dignity—all the polls show that—but using the experience of this minuscule minority of humans as the default reality for all of us—and teaching that as fact to children—is not an example of inclusion. It is an example of a well-meant untruth, imposed by fiat. Of course we've seen a reaction.

The "backlash" and "reaction" to which Goldberg and Sullivan were responding was the right exploding with a combination of real outrage and political glee. Mainstream Republicans seized on sex and gender as an election issue, using the combination of substance (Democrats want to do away with sex) and rhetoric (Democrats are crazy and dangerous) as a theme of their campaigns. The religious right, fated by their priors to play the long game, seized on sex and gender to win back lost ground. At least until the *Dobbs* decision came down—throwing abortion into the mix—this was a way to reach more than their base, including suburban moms who grew up as Title IX babies and black and Latino voters who tend to be traditional about sex. It was political cash which they parlayed into power, especially at the state level in conservative states.

But then, wherever the GOP took power, they overreached. Consistent with their underlying commitments, they enacted legislation that bans or severely limits abortion rights. (Life begins at conception and procreation is our highest value.) They adopted measures that prohibit gender-affirming care for minors and suggest that parents who would sign off on such care are child abusers. (No one is born in the wrong body, sex is binary, and doing anything that might risk their fertility is anathema.) And they banned books, lessons, and discussions that embrace or normalize gender diversity. (Consistent messaging and sticking to the gendered script are critical.) Lesbian, gay, bisexual, trans, queer, they were all on—or, in the case of gay people, back on—the block.

Florida's "Parental Rights in Education" law from 2022 is illustrative. It "prohibits classroom instruction by school personnel or third parties on sexual orientation or gender identity" throughout the curriculum. It was originally limited to "instruction" in kindergarten through the third grade, but as of this writing it's been updated to apply across the board—from K through 12; and it's being interpreted as banning classroom "discussion," not just "instruction," as well as student club activities—hence its nickname, "Don't Say Gay." The law's focus isn't just on the wacky curriculum reform efforts that properly concern Sullivan because they risk identifying gay kids as trans or undermining the parent-child relationship and the political deal between schools and parents. It also targets welfare-enhancing efforts by advocates who simply want LGBTQ kids to "see themselves reflected in the classroom and their non-LGBTQ+ peers" to "learn about LGBTQ+ communities."

When he signed the measure into law, Florida governor Ron DeSantis—who was said to be running for the GOP presidential nomination to the right of Trump—announced that "woke gender ideology" has no place in Florida. Doubling down after the Walt Disney Company, which has a substantial physical and commercial footprint in the state, vowed to work to defeat the measure, DeSantis replied: "I don't care what Hollywood says. I don't care what big corporations say. Here I stand. I'm not backing down." The Trevor Project, a support organization for LGBTQ kids and young adults in distress, responded: "The bill erases 'LGBTQ identity, history, and culture—as well as LGBTQ students themselves.'" Because dealing with suicidality is part of its work, when the Trevor Project uses the word *erase* you can be sure it's at least a double entendre.

Adolescence is an emotionally and physically turbulent time for all kids. As we saw in chapter 6, both boys and girls are suffering (different) epidemic rates of anxiety, depression, and other mood disorders in this period. We know that LGBTQ kids are in a high-risk category. We don't all agree that measures that disallow LGBTQ themes in school are the equivalent of "erasing" the kids themselves—there have always been lots of important things about us that our schools don't see—but most of us empathize with vulnerable kids and with parents of vulnerable kids. This happens to be the voice of a trans woman, Emily St. James, on Texas's prosecution of gender-affirming care, but it doesn't take being trans or even "for" this approach to gender dysphoria to get it, because she (too) makes a human—not a partisan—point:

> To me, a trans woman whose gender was harshly policed for almost all of her childhood, the definition of parents affirming their trans kids as "abuse" is positively Orwellian—a literal manifestation of "war is peace." The idea that children simply living their lives as themselves would be taken from families who loved and supported them and tossed into the foster care system (a potential outcome of the Texas measure) is a nightmare, and it's being sold under the guise of protecting children.

However we might approach the challenge of caring for our own trans child, picking on kids and parents is, by definition, risky political business. In a reverse of Biden on Mulvaney, the right has either been tone deaf or entirely uninterested in governing for (all) the people.

My sense is that each side is siloed to some extent. Our politics are so fractured in this moment that sitting down and having a conversation with the other side even to understand where they're coming from is usually not in the cards. Because of this, it's inevitable that sometimes they're misjudging people. But also, both are righteous and so at the end of the day, they actually don't care whether they're governing for everyone. The problem is, crusades that take away essential things from people are bound to create a political firestorm.

The Left's Assault on Free Expression

As I write, Florida's "Don't Say Gay" law is still in place for primary and secondary schools in the state; Republicans across the country are busy stripping school libraries of LGBTQ-themed titles; and right-leaning university regulators are using their budgetary authority to cull certain left-leaning programs. These moves have been inflammatory in the usual way of curriculum wars: people care a lot about what children are taught and how their tax dollars are spent in public institutions. In a pluralistic country like the United States, the battles always feature overreach, and Republicans clearly overreached in 2022 and 2023.

At the same time, the left's assault on free expression was undoubtedly an accelerant for the wider political firestorm. This assault is notable for its Soviet-like audacity and penetration. It's also been notably bizarre for its champions, which include the American Civil Liberties Union (ACLU), the premier historical defender of the speech rights of the most reviled members of our society. In other words, unlike the right's assault on the curriculum—business as usual for the GOP—the left's assault on free expression is revolutionary.

Free speech is a piece with our liberty, our equality, our competitiveness, and our commitment to the consent of the governed. Individuals who aren't free to speak their ideas or who don't have the words to convey their truth are neither equal nor consenting. Communities that block the free exchange of ideas inevitably sow the seeds of oppression and dissent. "Deplorables" and "listless vessels" will eventually make themselves known. We're pretty divided these days to the point where it's sometimes hard to say what we mean by "American" beyond shared geography, but there are (or were) still these common commitments because we understand (or understood) that they are necessary to our survival as a particular kind of political community.

The university's version of free speech is academic freedom. It's long been viewed as essential to allow faculty, operating within the bounds of their disciplines, to pursue unpopular ideas, research, and scholarship. The idea is that there should be a place in society where people work on issues from multiple points of view because (1) we don't know in the beginning who's going to be right (see Galileo), and (2) progress is often made when different points of view collide—idea one plus idea two equals unexpected value (see every day, all the time in the hard and applied sciences). This concept of the university—which is modeled for students who benefit from engagement with challenging ideas—justifies its existence in a way that's different from how advocacy groups justify theirs. Both are important, but they don't serve the same societal function.

Universities have substantive commitments too; they're not soulless vessels. Mine, for example, is committed to being an ethical institution that takes its history into account. It interprets this commitment to include (among other things) removing the artificial barriers that have kept women, people of color and from lower socioeconomic strata, and religious minorities from full citizenship status. The barriers exist because of Duke's history, and there's no ethical argument that supports maintaining their legacy. But holding these commitments doesn't settle the questions a scholar might have about the implications, or the institution's obligation not to interfere with their study. Duke economists Peter Arcidiacono and William A. "Sandy" Darity are both well known for their scholarship: Arcidiacono for work against affirmative action and Darity for work for slavery reparations. If only one or the other belonged here—felt supported and welcome—the university would be no different from an advocacy group.

Trans advocates on and off campus didn't violate these principles when they argued that trans people are who they say they are—that is, men or women. They violated them when they said: It erases us even to discuss this claim. It's settled, full stop. People who would speak about it should be shut down. And here are the words we insist you use and discard so that we can be comfortable. At that point, lots more people had skin in the game because it wasn't just about sex and gender anymore; it was now also about academic freedom and free speech.

That hasn't stopped them because, like religious conservatives, they're true believers. They reject the traditional model of the university and insist that academic freedom *should* be curtailed because "hate is not scholarship"—even when it's true to disciplinary norms. To them, the university *should* function as an advocacy group because good institutions are about justice. They also reject the traditional model for liberal societies—cultivation of groupthink and censorship of dissent are now legitimate strategies of social justice advocacy more generally. The ACLU, for example, still works on free speech issues but is increasingly concerned with things like its LGBTQ rights project. Consistent with progressive priorities in this period, the focus there appears to be on the T and the Q—issues of concern to trans and queer people—and on sex blindness across society. Controlling speech toward those ends is a key component of the strategy.

The playbook involves excluding dissenters from policymaking conversations, and that's basically everyone who isn't a full-on advocate for a sex-blind society. Among the excluded are most "cis-het" males—*cis* meaning that they're not trans, *het* meaning heterosexual—because they're toxic and have centuries of patriarchy to make up for. Also on the list are conservative women because they're pro-life captives of the patriarchy. And liberal women who don't accept their stance on sex even though we're usually allies in the quest for equitable polices. Instead we're called "transphobic" (because we insist on the material reality of sex) and "trans exclusionary radical feminist 'TERFs'" (because that insistence supposedly denies the existence of trans people). There's a special place in radical-left hell for liberal lesbians—the Ls in their own rainbow coalition—who disagree with their various positions.

Trans advocates do nothing to disguise their strategy and they aren't afraid to acknowledge their exclusionary practices. I can't tell you how many times I've heard that the question whether a trans woman should be eligible for a female space "isn't cis-women's business." Step aside. Step back. Step down. Even from a discussion about whether a space designed to secure our safety, our privacy, and our equality—*a space that exists for trans women to enter because of our distinct (gyn) biology*—should become something different. Where they have power, they're making policy for all of us with almost none of us in the room. It's hard not to see the irony: Joseph Bradley lives, he's just reading from a new script.

Here's the Trans Journalists Association *Style Guide* in 2023, putting my personal experience into published form:

> When reporting a story about trans issues, trans people should be interviewed and quoted as experts, not just subjects. Trans people are the experts on trans lives and experiences. Their voices should be centered in this coverage. A cis person will rarely have better insight. Trans people are also often the leaders in research on trans communities. There are cis experts worth quoting within specific

specialties; however, when you write about trans issues, do not include more experts who are cis than trans. *When covering trans issues, consider whether you need any cis voices in your stories.* (Emphasis mine.)

And here's how Fae Johnstone, an influential Canadian trans activist, put the matter of our silencing more generally, in a tweet in 2021:

> I actually do want a political environment in which TERFs are so vilified they don't dare speak their views publicly, let alone act on them. Shut. Them. Down.

Johnstone has since deleted that tweet and posted, "A strategy of the far right re: trans people is to force trans people out of public life by sending such a tsunami of hate our way that we come to loath [sic] ourselves." She's not wrong about this, as ultraconservatives like Matt Walsh make clear with their unbelievably cruel rants about trans women, for example, "[Y]ou are weird and artificial; you are manufactured and lifeless; you are unearthly and eerie. You are like some kind of human deepfake"—it goes on. It's hard, though, not to see that Walsh's is a version of the same dismissive, scorched-earth strategy Johnstone originally espoused.

Beyond these exclusionary practices, the radical left's playbook also includes censoring liberal dissent in the public sphere. This is accomplished not only by excommunicating us from their organizations and platforms, but also by convincing others that they should do the same. Sometimes calling us "transphobic" is all they need to do because that sounds like it must be a really bad thing, as though we're afraid of or hate trans people, when this couldn't be further from the truth. Some trans rights organizations publish lists of "transphobes" to make it easy for outsiders to check whether a person has been branded and so is on the "do not call" list. The scarlet letter is now a *T*. Here's the Trans

Journalists Association again, on feminists like me who aren't fully on board with their program:

> Some anti-trans rights groups and individuals use the euphemism "gender critical feminism" to describe their [fringe and] hateful ideology. They are also sometimes called "trans exclusionary radical feminists" or TERFs. This ideology should not be elevated in the press. When reporting on fringe groups and hate groups, instead of calling them TERFs or gender critical feminists, use language like transphobic, anti-trans, etc. *Avoid referring to anyone as a feminist when they are spreading anti-trans hate.* (Emphasis mine.)

Per this prescription, "feminist law professor Doriane Lambelet Coleman" becomes "transphobic law professor Doriane Lambelet Coleman." When replacing the descriptor *feminist* with *transphobic* isn't enough because they don't want us to appear at all, they actively lobby to have our names, our views, and even the facts on which our views are based erased from the discussion. I've watched this happen in the federal government, in the academy, and at the *Washington Post* and the *New York Times*, which shows how effective the tactic can be even in our most powerful institutions.

For example, I watched the *Times* for years avoid the operative facts that Caster Semenya's chromosomal complement is XY and that she has testes (not ovaries) that produce bioavailable testosterone in the male (not the female) range. When, in 2019, these facts were finally included in a story in which I was quoted on a different point, they only survived in the online edition for a couple of hours before being edited out. As I described in chapter 1, the misnomers "female with hyperandrogenism" and "female with naturally high testosterone" don't explain why her case on the merits came out as it did. Only the actual biology does. Yet this was a story about the decision in that case. Later that day, a commenter named Katie from Atlanta who had seen only the edited version responded to another commentor whose post included the relevant details, "Thank you for providing information that the NYT should have included in the story. It's almost like the NYT is deliberately obscuring the issue!"

Following the publication of the op-ed I described in the beginning of the chapter, the *Washington Post* struggled to cover the ongoing debate among women over the inclusion of trans women in women's sports. Best was that the debate wasn't covered if at all; if it's only on Fox News—if it's not also in the *Post* and the *Times*—*it's not happening*. When the reporting was pursued nevertheless—because it was hard to ignore that female sports icons like Martina Navratilova, Donna de Varona, Donna Lopiano, and Nancy Hogshead were in the fight—we learned that the ACLU was among the forces that sought for months to control the story, as though the paper were an arm of their movement.

For a time, the censorship of my name, my work, and the facts and science on which I rely was only an issue outside the academy. Mainly it was a press issue. That all changed in 2020.

At the invitation of First Amendment scholar Eugene Volokh and the Federalist Society, right before the country shut down for Covid, I gave a talk at the University of California, Los Angeles (UCLA) on Title IX and ways to accommodate transgender girls in girls' and women's sports. Leftist student organizations sent protestors with posters calling me a "TERF" and a "transphobe" and my work "transphobic"—along with "Fuck TERF" and "Transphobia is not welcome here"—to disrupt the event. They lined the hallway to the lecture room and then interrupted the beginning of my remarks to read their manifesto—the protest appears to have been led by a Marxist student organization—before clearing the lecture hall by threatening to report any students who stayed to listen to prospective employers: anyone who listens to a transphobe is a transphobe. They then photographed the mostly empty room and posted the image on Twitter with a comment to the effect that no one wants to hear what I have to say.

Back at Duke, a petition ostensibly developed by a group of under-graduate students and signed by members of the university community demanded that I be fired, or that the university dissociate itself from me, and that it publicly disavow my work. Then a group of students on the editorial board of one of the law school's journals demanded that a volume of essays I was co-editing called *Sex in Law* be scrapped or that a particular author be disinvited, and that its remaining authors be required to abide by an advocacy group's style guide. When the students didn't get their way—because they were asking that we violate academic freedom and professional norms—they resigned from their editorial posts. Never mind that we had curated a volume in which most of the authors were on their side of the issues. The point was that we shouldn't have platformed any who weren't—including me—because talking biology is hate and (again) "hate is not scholarship." All this made a splash in the *Chronicle of Higher Education* when the volume was finally published in 2022. My own essay, "Sex Neutrality," was the basis for this book.

There are, of course, much more famous instances of the radical left's attack on feminists who dissent from its sex-blind agenda. Bestsell-ing author J. K. Rowling's experience with this strategy tops the list. It started when Rowling challenged the trans movement's effort to dissoci-ate the word *woman* from pregnancy. She's also a fierce advocate for sex-based shelters, a cause she's long championed because she herself was a victim of domestic violence. Both positions—combined with her outsize cultural capital—made her a prime target. The story is well told in the podcast series *The Witch Trials of J. K. Rowling.*

Kathleen Stock's story is almost as well known. Stock, to whom I first introduced you in chapter 4, was the author my students specifi-cally targeted. Like Rowling, she is from the United Kingdom, where female feminists have thus far faced a much more physically threat-ening environment than we have in the States. She is an analytical philosopher and a liberal lesbian—all of this is relevant—whose scholarship was focused on feminism and fictions. She holds that women are adult human females, and that gender identity isn't part of sex. Using the tools of her discipline, she analyzed the claim "trans women are women" in her book *Material Girls: Why Reality Matters for Feminism* (2019). For daring to study a question progressive orthodoxy holds to be a settled matter, and for her conclusion that trans women are not women, Stock faced online abuse, her book was suppressed, her physical safety was threatened, and in the end her university concluded that it couldn't protect her. And so she was forced to leave her academic post. In 2021, shortly before that negotiated separation, she was awarded the Order of the British Empire by her government for "defending the rights of academics to discuss contentious ideas." Book sales surged. There are always costs to suppressing dissent, not least because people tend to want to know what the fuss is about.

In chapter 3, I described how the left's strategy includes changing the language. In addition to lists of transphobes, you've seen that they also produce style guides for institutions to use in their own communications if they don't want to risk being labeled transphobic themselves. Beyond requiring that women like me not be described as feminists or included in articles about trans issues, the guides typically would remove from the lexicon the words we need to talk about sex. These entries in the Transgender Journalists Association's 2023 *Style Guide*—especially when you put them together—are illustrative.

- "Avoid the terms 'biological gender,' 'biological sex,' 'biological woman,' 'biological female,' 'biological man,' or 'biological male.' These terms are inaccurate and often offensive."

- "Instead of 'born male' or 'born female,' which are inaccurate and considered offensive, use 'assigned male at birth' or 'assigned female at birth' (often abbreviated to AMAB and AFAB). You may also use 'raised as a boy' or 'raised as a girl' when appropriate."

- "'Male-bodied' and 'female-bodied' are inaccurate terms and are often considered offensive. Male and female bodies come in all shapes and sizes with various primary and secondary sex character- istics. *Instead use: assigned male/female at birth or raised as a boy/girl.*"

- Trans man: "A man who is trans. 'Trans man' is two words and trans is an adjective use to describe man. Making this one word is considered disrespectful and inaccurate, as it implies a trans man is not really a man."

Then there's the call to replace *vagina* with *bonus hole* or *front hole* and *preg- nant woman* with *pregnant people*—all ostensibly in support of trans men.

I hope I've been clear throughout the book that we should sup- port pregnant transgender men just as we support pregnant women. But if the estimates are correct, they comprise about 1 percent of the population and only a fraction of them will ever become pregnant. With respect to pregnancy and childbirth in particular, if prioritizing a tiny minority is necessary to the equity piece of the advocacy project, there's something wrong with the project. Especially when, as you can see, it's pretty easy to use the language in a way that erases no one.

It's useful to pause here to remember why we have language and then why vocabularies are what they are. "The Sami people," for ex- ample, "live in the northern tips of Scandinavia and Russia." They "use at least 180 words related to snow and ice."

This kind of linguistic exuberance should come as no surprise, ex- perts say, since languages evolve to suit the ideas and needs that are most crucial to the lives of their speakers. "These people need to know whether ice is fit to walk on or whether you will sink through it," says linguist Willem de Reuse at the University of North Texas. "It's a matter of life or death."

Classifying certain words as "unmentionable" is a way to say that what they signify is unimportant to the people who have the power to set the terms. When I was beginning my academic career, as part of a bigger project on the bounds of tolerance in pluralistic societies, I did work on the cultural defense to immigrant crime. One of the cultural collisions I studied was between American prosecutors and a set of Hmong refugee men who were charged with "rape." Defense counsel put it to the prosecutors that the men were engaging in a traditional practice called "marriage by capture," and it was said that the dialect they spoke didn't include a word for rape. If that was right, it doesn't mean rape didn't happen; it just means that the experience of the women and girls was unimportant to the men who controlled the lexicon. When American prosecutors decided not to judge that choice, they compounded the effects of the original erasure.

Using the word *gender* to mean *sex*, Caroline Criado Perez speaks to the more general point. "If you can't mark gender in any way," she says, "you can't 'correct' the hidden bias in a language by emphasizing 'women's presence in the world.' In short: because men go without saying, it matters when women literally can't be said at all."

It also makes people mad. In June 2023, NBC News reported an uproar at Johns Hopkins over the university's use of *non-men* instead of *women* to define the word *lesbian*. The headline read: "The university's online glossary of LGBTQ terms and identities defined the word 'lesbian' as a 'non-man attracted to non-men' before it was taken down." To this, Martina Navratilova, maybe the most famous lesbian of them all, tweeted:

Lesbian was literally the only word in English language that is not tied to man- as in male- feMALE, man-woMAN. And now lesbians are non men?!? Wtf?!?

Unreal . . . another example of erasure of women. Pathetic

It's a good thing she's too big to fail, as she speaks for a lot of us.

Some movement mantras are obviously right. Trans lives matter. Trans people belong. We should do what we can in the spaces we inhabit to make sure they know this.

Some version of the Equality Act should become law to ensure that this is the lived experience of trans people. Nobody should be denied access to goods, services, sports, work, or a seat at the community table because they are trans any more than I should because I'm female. Our governments promise, or should promise, not just life but also liberty and the opportunity to pursue happiness. Because we live in communities that are responsible for more than just these promises, we're never going to get everything we say we need to be free and fulfilled. But when local conditions stand as irrational barriers, they should go.

Everyone needs equality on the terms that matter for them. As I said in my testimony before the House Judiciary Committee in 2019, these measures may have made us all more free, but in fact not everyone needed the Emancipation Proclamation. Enslaved black people did. Not everyone needed the Violence against Women Act. Females did. Not everyone needs the Equality Act. Gay and trans people do. It sounds nice to say "we're all just human," but the fact is that ignoring relevant differences serves to perpetuate ongoing disparities. It deprives us of the ability to see and act on the truth that our circumstances aren't actually the same.

We know there are differences between transgender women and

females. We know there are differences between males and trans men. And we know that trans people who haven't taken cross-sex hormones or had surgery to feminize or masculinize their physical bodies are different from trans people who have. Everybody matters, everybody belongs, but we're not the same.

All of these differences relate to sex. Some, but not all, trans people take the position that it's offensive and disrespectful to notice any of it. They will say that I shouldn't have written the previous paragraph, or this book. "Hate is not scholarship," goes the refrain. But it can't be this way. Sex matters fundamentally to most of us. *We can carefully hold two truths—sex and gender—in our hands at once.*

I'll use myself again as an example. I don't have a gender identity. Please don't tell me that I do; since this is about my own inner sense of myself, that's not your place. I do have a sex. I know this because I can see it, whether it's with my own eyes or on scans and in blood test and urine sample results. I also know from years of experience that my sex is female, that it's not just a part—a uterus, for example—or even set of parts—for example, XX, ovaries, and testosterone levels in the female range—but rather a whole form, an integrated system, and that to a large extent that form and system have defined my life. That is, I've had a particular set of physiological and social experiences because my body developed as it did, according to my XX blueprint. The words I've always used, because they're the ones in our lexicon for all of this, are *woman, girl,* and *female.* I connect these up to sex, not to gender. I'm not saying that you can't do differently, but I am saying that to the extent my experience is multiplied across the population—as it surely is—you shouldn't try to redefine me or anyone else according to the different trait that defines you. I'm for reciprocity, not erasure.

9

A Commonsense Approach

I WAS BORN IN LAUSANNE, Switzerland, in 1961. The political tumult of the 1960s and early '70s in the United States was the landscape of my childhood. I was in public kindergarten and elementary school in New York City through the end of the 1960s when Dad got his draft notice and conscientious objector status, my brother Tokya was born, Martin Luther King Jr. and Bobby Kennedy were assassinated, and my sister Lalou arrived.

In chapter 5, I told you a bit about my Swiss seamstress mom, Bluette, at the end of her life. She was an adventurer, always looking to escape convention and to be challenged by the richness of the world. She was also a *vive la différence* feminist, which, in retrospect, was probably the basis of her strength. Part of her vocational training centered on the use of haute couture to celebrate the beauty of the female body. There was a particular dress from when I was pregnant with my second son that captures all of this. She made it out of a piece of raw silk that she got on a trip to Bangkok ten years before I was born and—well before this was the style—it showcased my belly. She met my dad, Hank, on a later adventure to California, at a party on a houseboat in Sausalito.

Dad was born in New York City, into a black multigenerational family of civil rights advocates—in addition to Ellen and William Craft, the Jefferson-Hemingses and William Monroe Trotter were ascendants. Mom and Dad drove back east in his old Ford pickup—he was a carpenter—camping along the way because this was before a black man and a white woman could check into a motel together. But also, this was their way. Neither went to college but both read voraciously and had portable trades that almost always paid the bills.

In the 1960s, we lived first in the Puerto Rican section of Manhattan's Lower East Side ghetto and then, beginning at the end of the decade, at the corner of Haight and Ashbury Streets in San Francisco. "We weren't hippies," Mom would later say. "We were bohèmes." She's on film in a French documentary from the late 1960s called *Les Enfants de la Ville* explaining why we lived on the Lower East Side, where they also filmed me running through streets that look in retrospect like an apocalyptic landscape. "*Il y a plusieures manières de construire un être humain*," she said. There are different ways of shaping a human being.

Depending on the day, Dad was more Malcolm than Martin. Often music tells the story of a family and a life. Ours was Aretha Franklin's "Respect." The Brandenburg Concertos. Nina Simone's "Strange Fruit." Vivaldi's *Four Seasons*. And at the end, Marvin Gaye's "What's Going On." Dad died by suicide just before my eleventh birthday, but I still have his original vinyl along with a black-and-white photograph of me on his shoulders at a freedom march in Central Park and lots of searing memories. One is of how Mom picked herself up out of that dark time and her bohemian life, summoned her inner Swiss pragmatism, and took us back to Europe.

At the end of February 1972, she was suddenly a single mother of three with only half the income it took to be all right. The conventional life she had eschewed when she was younger drew her back to the continent and to her own mother, Lily. My grandmother had also been widowed young, in her case with six children in the middle of World War II. She wasn't complicated and she loved us beyond words, which was exactly what we needed in that moment. Switzerland would never be right for Mom, though, and so several months later, she took us back across the Atlantic and finished raising us in Woodstock, New York. There, Mom said, it was possible to do that well—by which she meant unconventionally and without feeling poor.

Woodstock is a village nestled in the heart of the Catskill Mountains a couple of hours north of New York City. It's most famous as the namesake for a multi-day festival held nearby in 1969 that the National Park Service describes as "the definitive expression of the musical, cultural, and political idealism of the 1960s" and "a watershed event in the transformation of American culture." The village was founded in the nineteenth century as an artist colony, and it remains that—and politically idealistic—to this day. It's partly from their own Woodstock summers that my sons learned that art is one of the ways to see the world.

I tell you all of this because you should know my priors too. One of Mom's refrains was—cue the strong French accent—"Life is hard." It just is. Adding beauty and joy whenever you can, even in the smallest ways, makes it better. And America has never been just a place to me. It has also always been an idea: an idea of freedom along multiple dimensions and of people from different places, personal orientations, and cultural habits not just tolerating each other—that's the bare minimum for a political community to survive—but doing what the French call le vivre ensemble. My great friends Géraldine and Stephen Smith, who are French and write political books about Africa and France (and in Géraldine's case, about America too), translate this as "living together respectfully." My version, which I was raised to think of as the American version, is more sanguine. The curry metaphor is overwrought, but because it's been my life, I believe in its possibilities. I also know how hard this America is to curate and that the project needs constant tending and care. When our politics are fractured, as they are these days, it's good to remember that we can usefully toggle between toleration and *le vivre ensemble*, and to work purposefully on that.

Among the more specific things that flow from my parents and my upbringing—the good and the bad parts—is that each of us matters and we all have a right to dignity and respect. That we're all connected even as we might prefer it otherwise sometimes, and so it's always valuable to cultivate the connections when we can. That race and sex aren't the same even as we've been segregated, and either privileged or subordinated, because of both. That sex is complicated but that it's always there and always matters. That fathers and men are important even when they're gone. That mothers and women are special in part because they stay. That we need to see our kids not just as projections of our own visions; we need also to be fiduciaries of theirs. That it's callous to impose your beliefs on others who don't share them and for whom the impositions make an already hard life harder. That ideas can be intellectually engaging and important to pursue but that the rules for society must work for real people in real life or it's inevitable that they'll be unjust.

Another way to put this last point is that I prefer pragmatic, commonsense approaches to real-world problems. I think of common sense as a shared view—the word *common* meaning "held in common"—which is both fact-based and caring, or at least harm-minimizing, for individuals and their communities. When someone asks, "What's the common-sense approach to this?" to me, the answer is "The thing or things that will work for real people in their actual lives."

Common sense doesn't ignore our aspirations; it generally supports them. If we want to reduce the rate of sexual violence—which is undoubtedly an aspiration—a commonsense approach is to notice the fact that it's mostly a male-on-female phenomenon and so also to focus on raising good men. Masculinity isn't inherently toxic, but it does need love and guidance. If we want America to work as a liberal, democratic, pluralistic society—given the course of human history, that's a huge aspiration!—it's common sense to check our ideas for compatibility with this political goal: Will they work for people among us who have a fundamentally different worldview; and if not immediately, then how can we bring them along? As we learned from the work it took to realize the promises in *Brown v. Board of Education* (1954) and *Frontiero v. Richardson* (1973), and the work that's ongoing on the promise of *Obergefell v. Hodges* (2015), successful pluralism is cultivated. But how exciting for us when we do that well! As I write this, I'm thinking about Mayor Pete in 2020, campaigning for the presidency in conventional places alongside his husband, Chasten. It's still fragile but, like *Obergefell* itself, the Buttigiegs' willingness not just to walk that ground but also to build bridges has done so much positive work for all of us.

On that point, in a country like ours, it's not only interesting—a piece of the richness of the world—it's also common sense to talk to people with whom you disagree, to understand where they're coming from, and to share your side. Abortion is the essential example. Many of my lefty friends assume that the only thing that could explain why conservative women would agree with the right's position on abortion is that they're captives of their husbands and ultimately of the patriarchy. Because of this, they have no interest in what these women have to say. The truth is much more challenging—they're as wrong that conservative women lack agency as conservatives are wrong that liberals are lost or barbaric. But we'll only really understand this if we talk to each other. ·

My former student Denise Harle was for a time the director of the Center for Life at the Alliance Defending Freedom and, in that role, advocated for pro-life laws. She put it to me this way when I asked her about the charge that conservative women lack agency:

If you ask a woman why she had an abortion, most times it becomes apparent that she didn't feel she had any other option. It's cruel to pit women's interests against their children, suggesting a mutually exclusive "choice" between motherhood on the one hand, and a successful career or happy life on the other. Women can find freedom in knowing that they have real support from their partners, communities, and employers when faced with a pregnancy—planned or unplanned. Real empowerment is celebrating that women have a unique ability to give life.

Before you pounce, consider that the idea contained in this last sentence lies at the heart of every matriarchy everywhere, and that it's often shared by liberal women in places that do a better job supporting mothers. Denise's second-to-last sentence doesn't work for the pragmatist in me for a related reason: too many women don't have support from partners, communities, or employers that facilitate the fulfillment and freedom she describes. Where I'm from, abortion and the pill are central not just to our ideas about liberty and equality but also, *even mainly*, to our ability to live decent lives, and to raise children—if that's what we want to do—without feeling or being poor.

We're never going to agree on everything, and that's not the point anyway. But opening ourselves up to the conversation grows understanding and empathy and makes collaboration on shared commitments possible—for example on the ideas of fulfillment and freedom and on the basics that are access to contraception and support for mothers and children. Leaders on the left and the right on issues of sex and gender do us all a disservice when they brand their opponents as heathens and hate groups. My guess is that most of us are eager and willing to cultivate common ground.

There's utility in the push and pull of divergent ideas and, as part of this dialectical process, each side adds value. It's wrong, though, for either to try to govern in ways that aren't true to the facts of real people's lives and to our collective commitments. Which brings me to some common sense on the issues I've raised throughout this book.

Sex Is Binary

As the renowned evolutionary biologist Richard Dawkins explains it, while most attributes we have occasion to consider—"Tall vs short, fat vs thin, strong vs weak, fast vs slow, old vs young, drunk vs sober, safe vs unsafe, even guilty vs not guilty"—are on a spectrum, "Sex is a true binary." We have either a Y chromosome and a body that "masculinizes" with lifelong effects because of that base, or we have X chromosomes and a body that "feminizes" with lifelong effects because of this different base. Both types of bodies are integrated anatomical and physiological systems designed in the first instance so that we can reproduce. *Sex* is the word that signifies these reproductive systems.

As for genuine intersex people—meaning people who have both ovarian and testicular tissue—Dawkins adds that they're "way too rare to challenge the statement that sex is binary. There are two sexes in mammals, and that's that." As I explained in chapter 1, "too rare" translates to about five hundred total cases in recorded history.

Almost everyone who is described as "intersex" or as having a "disorder of sex development" (DSD) is simply atypical in some respect for their male or their female sex. Caster Semenya's chromosomal complement, for example, is XY; she has testicles; they produce bioavailable testosterone in the normal male range; and her body masculinized through puberty—which is why, for a decade, she dominated the women's 800 meters on the global stage. Neither her particular DSD nor the decision to "assign" her the female sex at birth changes any of this male biology. They also don't change the fact that she's a valuable human exactly as she is. But for the mismatch with the female category in elite sport, we can responsibly guess from her public biography that she would be a vibrant member of her communities, presenting as male, who happened to have fertility issues.

Some academics and advocates insist that sex is neither binary nor about reproduction and that DSD prove their points. Actually, DSD prove the opposite. Atypical sexual development may make knowing a person's sex more complicated than the standard genital check at birth, but the diagnostic process shows us that it's always there. Doctors aren't "assigning" sex at birth, they're recording it—like a meteorologist records the rain. Atypical sexual development also means that making babies can be complicated, but with support, it's often possible.

There's simply no way to pretend sex isn't binary. But we must also acknowledge that many people, especially kids and younger adults, are experimenting with gender in ways that don't cause harm. We can't all embrace this new gender diversity—I do but I get that it won't work for everyone—but we can all be polite about it. *We can hold two thoughts at once and carefully; we can affirm gender without engaging in a pretense about sex.*

Sex Is Different from Gender

Although sex and gender are sometimes conflated, the ideas they signify are importantly different. Our bodies and how they develop and age given our chromosomes and endocrine systems is one thing. That's sex. What our cultures do with our two physical forms, the social constructions that are based on our sex, and then how we conceive of and express ourselves, is another. That's gender.

Sex and gender are both important as we go about the business of living. It's essential that we have the words and permission to talk about both—that we can say "sex" and "gender," that their definitions are clear, and that the two are distinguished in relevant circumstances. We saw in chapter 5 that the most consistent public plea on this point comes from the medical and research communities, which need to be rigorous about the distinction for the promise of their work to be realized. They can help us to be sex-smart only if *sex* and *gender* aren't muddled. But the words are essential beyond that.

I mentioned in chapter 8 that some people and institutions challenge the use of the term *biological sex*. The Associated Press, for example, takes the position that it shouldn't be used because it's redundant and because it's offensive to trans people. This second point comes from the trans rights organizations and their journalists' style guide. The AP is right that it's redundant to add the adjective *biological* to *sex* if we consistently use the word *sex* as biology and distinguish it from *gender*. But it's not redundant if we conflate the two, which we do a lot. Because of this, when we want to clarify that we're talking about the biology, we need the ability to say so.

It's gotten especially important of late as you'll sometimes hear someone say, for example, about a transgender boy—a female child who identifies as male—"his sex is male." There's lots going on here and the muddle is part of the strategy for advocates. But common sense asks us to be as clear as possible and the easiest way to avoid communication errors is to distinguish sex from gender. The suggestion that talking about biology is "anti-trans" is only right if the aim is indoctrination rather than information.

Lots of trans people will tell you that this indoctrination is rubbish: that the biology matters to them too and that they're not asking us to pretend otherwise. But what's clearly not rubbish is that trans people with dysphoria—especially young trans people—can be hurt by our attention to sex. This isn't hard to understand. Just change the facts to make them work for you: You're in a situation where someone asks you about something that stirs up hard feelings. Suicide and sexual assault do it for me every time, which is why I'm triggered by the suggestion that I can't talk about sex without risking the suicides of trans kids. That sports went a long way to saving me in adolescence and that I'm often talking about sex in that context just dials up already-elevated cortisol levels. It's unrealistic to think that we can go through life without being upset. I have no illusions about that. It's also wrong to take important topics off the table. But we can try to be kind to others and to ourselves as we go.

I don't avoid sexual assault when I teach the tort of battery—it's not only about bar fights among men, and every student in the room should be able to make the case for women too. But I also don't cold-call anyone that day. If a student lets me know in advance that their parent died by gunfire and we're doing a related case, they come to class, they learn the material, but they don't have to present it. We can do this together.

Sex and Gender Are Both Good!

The *David* and the *Venus de Milo* are no more indecent than our bodies. If you see them as pornographic, you've got (having) sex on the brain. That's not about them, it's about you. That we all recognize these artworks is proof of a shared sense that they're gorgeous renditions of the natural human form. It's utterly irrelevant to our appreciation for these timeless treasures that the model for the *David* was gay or bisexual (as the archives establish) and that the model for the *Venus* could well have been nonbinary (there are no archives here but it's a statistical possibility). Sex is independently good.

Gender is different from sex, and it's also good. Cultural expressions of masculinity and femininity—clothing, jewelry, cosmetics, hair, and so on—can be beautiful whether they're gender-conforming or gender-bending. For the former, check out Naomi Campbell's "Most Iconic Photographs" in British *Vogue*. For the latter, think about Timothée Chalamet's "gender-fluid" masculinity, or a "girl dad" painting his nails blue too—Kobe Bryant made this term famous as a way to say he was all in as a father of girls. Gender bending is sometimes also the embrace and expression of a person's inner sense of themselves that is naturally different from their sex. Here think about a transgender person who isn't, and then is, out as trans. When they're free to be out, when the people around them are neutral (at least) or welcoming (at best), that's joyous for them and joy adds beauty to the world.

A dose of grace all around: it's not wrong to notice sex because it's adaptive. Remember Caster Semenya and Violet Raseboya's story, which was, just for a moment, about privacy and safety before it became about beauty and love—and in the last few years they've added marriage and parenthood. It's also not wrong to notice incongruity between sex and gender because gender expression is typically a reliable proxy for sex. You may feel cognitive dissonance if you see a male-bodied person expressing themselves in a culturally feminine way, but once you get past that reaction, once you've got your bearings, you have the choice to resolve it respectfully. Unless there's something in the way of that choice, take it; then let them know they're safe too. It's scary to think you're scaring someone.

Bathrooms—and Other Sex-Based Spaces— Are Important to Get Right

To make sense of debates about sex-based spaces like bathrooms, I recommend you focus on these three questions:

- Why is the space sex-based?

- Are these good reasons or should the space be unisex—like modern law school libraries?

- If the reasons are good, what are the pros and cons of making an exception for trans people who want to be included based on their gender identity, not their sex?

From chapter 7, you know this is basically the way we decide whether sex discrimination is constitutional. My guess is that it's also the way we think through most of our own choices.

We typically agree on two reasons for separating the sexes in public spaces: safety and privacy. Both are tied to the differences between the male and female body—including their different behaviors—and to the value we place on the existence of spaces where people whose bodies are one way don't expose themselves to people whose bodies are another way. Other associational reasons—religious, social, educational, and so on—to separate the sexes are also important; I want to be clear that I'm not discounting those, just recognizing that they aren't as uniformly agreed-to as safety and privacy.

Privacy is culturally relative, meaning that different cultures have different privacy norms, but where they exist, they're important to individuals and their societies. Physical privacy is interesting because in one way or another, it's an almost universal norm, which suggests it may be innate and adaptive. Regardless, we don't typically expose our "private parts"—directly or indirectly—to non-intimate members of the opposite sex. I was routinely drug tested when I was a competitive athlete. Wherever I was in the world, if a man was in charge of the testing there was always a woman assigned to accompany me when I provided my sample because that required watching me urinate. Female athletes accept this because it protects the integrity of the competition and we've been naked in open locker rooms with other women all our lives. No one asks what our gender identity is because that is irrelevant. It's about the body.

Safety isn't culturally relative. As the UN report I described in chapter 6 makes clear, females around the world are vulnerable to male sexual violence in places where they go to relieve themselves. Although some regions of the world are safer than others, it's everywhere an issue and our antennae are properly always up. If women and girls don't feel safe, we don't go or else we're wary when we do. I get that trans advocates don't want us to think about trans women and trans girls as males, especially not as predatory males; but their casual dismissal of our over-arching concern—and of our instinctive adaptive response—isn't helpful. Women and girls weren't safe in public bathrooms before they were sex segregated. Those who defend separate-sex restrooms make a mistake when they avoid this issue.

Both safety and privacy are important to people, including to trans-gender people. In fact, as far as I can tell, outside of very privileged places where there's ample money for a private restroom, sink, and mir-ror for everyone, it's mostly some nonbinary people who would take down the walls between "men's" and "women's" restrooms. This means that for most of us, unisex restrooms don't make much sense, and we're left to focus on the costs and benefits of making an exception for trans-gender people who would be included in one or the other based on their gender identity and not their sex.

This issue came to the fore for me in 2016, when my home state of North Carolina passed a statute, known by its bill number HB2, re-quiring trans people to use bathrooms that correspond to their sex, not their gender identity. At the height of that controversy, I wrote an op-ed opposing the legislation. The way I think about it is that I might have a momentary reaction when I see a transgender woman in a bathroom set aside for females, but I realize quickly that the person is transgender, and trust returns. That is, my instinctive reaction is followed by the ra-tional understanding that the risks are low that the person is there to as-sault me rather than simply to relieve themselves. This to me is settling. I'm personally comfortable balancing whatever remaining concern I might have against the need for trans women and girls to have a safe and private place to go too. If they're phenotypically female or presenting as female, that's not the boys' or men's room. My own cost-benefit calculus is complete at that point. It's not always first nature but it is second.

Not all women and girls feel the way I do and given the two main reasons the space exists as sex based, we need to respect that. For some, trust doesn't return as easily or at all. Like me, they feel strongly about the fact that our protective instincts are adaptive, but they disagree that they should consider adjusting that reaction. They don't think we should have to sideline our physical safety and privacy interests. That many societies are finally reaching the point where females can define for themselves the boundaries of their comfort levels around males—and that we can now say no to unwanted exposures—supports this important political claim.

Kim Jones is the founder of the Independent Council on Women's Sports and the ferocious mom of a Yale University swimmer who competed against the University of Pennsylvania's Lia Thomas. Jones is smart and passionate on this score. She told me that when Thomas came out as trans and moved from the men's team to the women's team, Penn made clear that it supported her unrestricted use of the women's locker room, that the females on the team could change elsewhere if they weren't comfortable undressing in the same space, and it offered them counseling if they couldn't accept the situation. Harvard followed suit, suggesting to its women that if they weren't supportive when the teams met up, they could be putting Thomas's life at risk; and Yale did a version of the same. As Jones put it to me,

This left the young women (or swimmers) with nowhere to turn. The schools pulled them into mandatory meetings where the swimmers were intimidated and silenced. A clear message was laid out that the administration, all the way up to the top of the Ivy League, was standing with Thomas and that the girls' concerns regarding locker rooms and competition would fall on deaf ears. Some swimmers found hallway bathrooms, toilet stalls, and closets or practiced changing under towels, but many of the girls were resigned to believe they simply didn't have a choice. Who was left to help them?

We must find a way to support trans women, including in their also-legitimate safety and privacy concerns, that doesn't involve sidelining, silencing, and gaslighting females. The notion that an institution in this day and age would tell females that there's something wrong with *them* if they don't consent to being exposed to male genitals in a female space is just wrong. Our work here isn't nearly done.

Gender-Affirming Treatment Should Be Evidence-Based and Compassionate

The fight over whether trans kids should be able to get gender-affirming medical care is difficult almost beyond compare because it's about kids in distress, parents who are trying to figure out what to do when there's a lot that even pediatric medical experts don't yet know, and outsiders—elected officials and advocates on the right and the left—who are politicizing their circumstances. Medical standards of care must be evidence-based. Patients and their families must be treated with respect and compassion. Neither should be political footballs. In this context, being evidence-based and compassionate means being clear-eyed about what we know and don't know about pediatric trans medicine and applying that knowledge—just as we do in analogous circumstances—with an eye toward what's probably best for the individual child.

When we're clear-eyed, we see that puberty blockers, cross-sex hormones, and cross-sex surgery are all major interventions in the normal physiology of the healthy body. We see that even if puberty blockers merely pause puberty for a year or two—a claim that's hotly contested—it's still a big deal not to develop neurologically, physically, and physiologically at one's natural pace and consistently with one's peers. We see that even if cross-sex hormones fully replace the child's natural endocrine system—which isn't clear because we don't yet understand all that our endocrine system does, especially neurologically—they affect the development of our genitalia, which are critical to healthy sexual function. It's not standard pediatric practice to do gender-affirming top surgery until a trans person is at least sixteen—and waiting to eighteen is standard for genital surgery; but the removal of any healthy body part, not least breasts, is always a matter of physical and ethical significance. Finally, we see that the treatment may also complicate fertility, which has always been factored into medical decision-making for kids. One of the first law and medicine projects I worked on years ago was the development of a protocol for pediatric oncology patients to harvest and cryopreserve their gametes for future use should the treatment be successful.

Because doctors, parents, and the state are fiduciaries, as with any other treatment, before we proceed, we should know that the costs are probably outweighed by the benefits. Taking this approach, it's common sense that a child who experiences gender dysphoria in early childhood— who we know has never been able to reconcile what's between their legs with who they understand themselves to be—should be supported, including medically if that makes sense for them, when we have the means to do so. Because we can be quite sure that their sense of themselves hasn't been influenced by anything other than atypical neurological development either in utero or in mini-puberty—most of these "early onset" kids are male—and because we know from generations of trans women that it can be deeply distressing to experience "the wrong puberty" and then to go through their lives in physically masculine frames, a parent with their doctor is probably being a good fiduciary if they choose gender-affirming treatment starting at the onset of puberty. The "gender-affirming" model of care—which replaced "wait and see"—was developed by the Dutch for this cohort of kids.

To me it's also common sense that a teenager who is eager to curate a preferred aesthetic shouldn't get an adult to sign the consent form or to provide treatment. That adults can make this choice for themselves goes without saying, but kids in the throes of puberty aren't treated as fully autonomous about anything important precisely because their brains are a hormonal mess, they tend to be short-term thinkers, and they don't have much life experience. It's especially important for the adults in the room to be responsible today, given how female puberty is starting earlier and earlier, and the extent to which girls are being bombarded with social messaging about how they're supposed to look to get by. It's healthy to support kids in their gender expression and great to see them experiment with impermanent artifacts. But decreeing that a child whose gender expression is different from their sex is a trans kid who has a right to hormones is irresponsible both because it risks misidentifying gay kids as trans, and because it requires factoring into the treatment decision unestablished ideals about gender that the child may not share when she grows up.

As I write, the fight is where it should be, over kids—mostly females—who come out for the first time in early to mid-adolescence and who often have other issues. To contrast them to the original cohort, they're sometimes described as having "late-onset" dysphoria. Depending on who you talk to, they either always knew they were trans but didn't say so, or their sense of themselves as trans emerged along with puberty. Given what we know and don't know about this second cohort—which is a lot—it seems equally problematic to assume that they do as it is that they don't need blockers and cross-sex hormones.

On the one hand, as we saw in chapters 1 and 5, female brains develop differently from male brains. Then, unlike males who get signals early on that they shouldn't be feminine, females generally aren't discouraged from behaving like tomboys, and so they may not realize until puberty how consequential that developmental stage will be. Finally, there are families and communities in which coming out as trans earlier isn't easy or even possible.

On the other hand, unlike the early-onset cohort, there's ample evidence that the late-onset group has been influenced by other people and the culture—including about the buzzwords that can open the door to blockers and hormones, like "Looking back, it's clear we always knew." That we're talking mostly about females who are already more likely to suffer from anxiety and depression *exactly around the facts of being a female-bodied person* means that it's especially important that we're evidence-based and cautious about each individual diagnosis. Let's not check our common sense, our fiduciary obligations, or standard medical ethics at the door to the endocrinologist's office.

Two final points:

The right is wrong—factually and ethically—when it acts as though dysphoria isn't a real condition, as though trans kids aren't on the range of normal human presentations, and as though it's somehow justified to be cruel to them and to their parents as they navigate their way through their already-difficult circumstances. It makes no sense—and this is usually common ground—to pathologize kids and demonize their parents, and thereby to add to their distress, when we know that if we treat them with care and respect this will lessen that distress and contribute to the chance they can live fulfilled lives. Why wouldn't we do that?

The left is wrong—factually and ethically—when it throws "suicide" and "suicidality" around like candy in its effort to prevent reasoned discussion in individual cases and about pediatric policy in general. Medical practitioners shouldn't be allowed to recite the movement mantra "You either have a trans kid or a dead kid" to get parents to sign off on blockers and cross-sex hormones, as though gender dysphoria were the equivalent of leukemia and these treatments the equivalent of oncology drugs. No one wants children dying and trans kids apparently do think about suicide more than many of their peers. But suicidal thoughts aren't the same as suicide and there's a dearth of evidence about the latter. Even as they can be enormously helpful for some kids, blockers and cross-sex hormones do have risks and they aren't the only treatments for anxiety and depression. Parents should be fully and honestly educated

249

about the risk-benefit calculus of treatment so that they can make good medical decisions with or for their children.

Elite Sports Should Be Classified by Sex, but Gender Matters in School Sports

What we call "sports" is really a set of related institutions. Their common feature, what makes them all "sports," is that they're about organized physical activity. Sometimes, but not always, a piece of this is competition. Kids, adults, and elite professionals may all play a game called "basketball," but the reasons they do this are quite different. The rules of the game may even be different as a result. It makes sense to take all of this into account when we're deciding who gets to play where. My sense is that people understand this but we're fighting about it as though we don't because sports has become a proxy for the political fight over sex itself. To do the right thing here too, we need to let the politics go.

I'll start where we all start, as kids on playgrounds—and in gyms, on fields, in pools, and on ponds. If you're thinking elite sports from the get-go, these are development opportunities: Team USA is born here. But if you're not—and mostly we're not—these are opportunities for fun, for growing physical literacy, for health and well-being, for discovering camaraderie, learning sportsmanship and how to follow rules, and so on. Whether it's in or out of school, especially early on, boys and girls typically play together—sport is coed—because competition isn't the main point. The biologically driven performance gap between males and females is there from the beginning, but it's not yet so obvious, and it's good for both to learn to be physical with each other in a respectful way. Altogether, these are the reasons public schools fund kids' sports programs.

Girls and boys are usually separated around the start of puberty, the developmental period when we mature sexually and the performance gap explodes. That gap—between boys as a group and girls as a group—goes from about 5 percent prepuberty to between 10 and 30 percent postpuberty, depending on the sport and event. As this happens, the setting also becomes more competitive and winning becomes an ever more important aspect of the activity. This is when students start to be included and excluded from teams and play based on their physical abilities.

What all of this boils down to for me is that so long as the activity isn't competitive, the performance gap between males and females isn't so obvious, and it's not a contact sport like wrestling or tackle football where physical safety takes precedence, trans kids should be allowed to play with the groups that match their gender identities if they want to. Where the predominant theme is "everyone can play," it's community-building and kind to be inclusive. *We're making an accommodation for gender in a sex category in circumstances where inclusion does no harm.*

The harder call is whether trans girls who've experienced male puberty should be allowed to play on girls' teams that are more competitive. I've done good battle with experienced friends on both sides of this most contested issue, and they make important points, but in the end, I come out in favor of inclusion until competition and selectivity become the predominant themes. Based on our arguments—which you'll see built into my explanation—here's why:

It's true that I learned that a female's body is strong by the opportunity to race only against other girls. (By age thirteen or so, the boys had grown their turbo boosters, and because of that, you could be the best female in the state but go down every day in practice to mediocre or second-tier males.) It's also true that I learned how to win on the big stage by winning on little ones first, so those regular season dual meets on small-town tracks were important confidence builders. (It's an adage in elite sport that you learn to win by winning.) And yes, including trans girls who've experienced male puberty on girls' teams and in girls' competition sends the message that being a "good girl" sometimes means stepping back for others when "good boys" aren't asked to do the same. But if the program is still mainly about physical education and healthy fun, I come out in favor of respecting that. Although we celebrate winners, we don't fund school sports—as opposed to amateur and club sports—to give singular athletes these opportunities.

My approach is more complicated than the alternatives: "you're in" based on your gender identity or "you're out" based on your sex. But so long as the numbers of trans girls are small it should be easy to manage. How you define transgender is key here, as is making sure the responsible adults don't gaslight the girls with false statements about the biology.

I can't emphasize this last point enough. We have girls' teams for trans girls to join because male physical advantage—which they undoubtedly have—isn't comparable to anything else. Katie Ledecky is said to be "better at swimming than anyone is at anything." My friend and occasional coauthor, Mike Joyner, is a global expert in human performance. He adds that among the athletes he and his colleagues can measure and compare, it's Ledecky and Secretariat at the top by a wide margin. Yet—like Navratilova's—we wouldn't know Ledecky's name if she'd had to compete against males, whatever their gender identities. Educators must tell the truth about the goals of the inclusion exception, so that the girls don't feel sidelined and misled, and so that they can be honestly supportive.

What's not a hard call for me is that—however they identify—a male-bodied kid shouldn't be the girls' state champion. There are lots of reasons a girl might be on the team but sit out a competition, most of which have no opprobrium attached. To me, male performance advantage—the only thing more than 99 percent of males have but no female can begin to access without doping—is among them. Invitationals, events with knockout rounds, and postseason play should also be restricted by sex because the institutional focus is now competition.

The same holds for elite sport, where money and national team spots are on the line. It bears repeating the bottom line from chapter 5: the exclusion of males is the female category's raison d'être. If cross-sex hormones could eliminate the constructive effects of the Y chromosome and male puberty, it would make sense to consider an exception for trans women who are on them, but they don't. Australian American sports scientist Sandra Hunter is known for her work on sex differences, including on the range of "legacy" or "retained" advantages following gender-affirming hormone treatment. Among the most significant, she says, are differences in skeletal mass and muscle fibers, differences in heart and lung size, and different stature and bone length. This doesn't count other important sex differences in the foot, in joints, in tendons and ligaments—the list goes on. The reason that trans men (like Cal Calamia) aren't of concern in the male category as trans women (like Lia Thomas), and athletes with XY DSD (like Caster Semenya) are in the female category, is that no amount of testosterone supplementation can replicate the constructive effects of the Y chromosome and male puberty. These aren't myths and stereotypes, they're facts. It's bad for women and girls when we pretend otherwise.

As we work through these challenges, it's important to hold on to the difference between sex and gender. Competitive sport separates us by sex for good reasons, but it also welcomes—or should welcome—gender diversity within sex categories, just as it does—or should do—within age categories. Athletes have always had gender identities all along the spectrum. Caitlyn was inside Bruce when Bruce won gold in the decathlon at the 1976 Olympic Games. Nikki Hiltz, the new American record holder in the women's mile, identifies as transgender nonbinary and uses *they/them* pronouns. They belong in the female category because they're female; they're celebrated as a member of Team USA because they're a great athlete in their sex class; and they're embraced on the international circuit because they're a special human being.

If We're Honest, Pronouns Are Complicated

As I've done with Nikki Hiltz, throughout this book I've used named trans people's preferred pronouns. I haven't done this to be politically correct. I've done it because in these instances, the two functions of gendered pronouns as I understand them have been served. The first is to convey information about a person's sex. For example, your boss says: "I want you to have dinner with Terry. [She or He] is an important potential client." Remember the hardwiring I talked about in chapter 6. Because I'll factor my sense of Terry's sex into my decisions about the assignment—whether it's to see the bigger picture, check myself, and dismiss its relevance, or to book a restaurant with a bustling, open floor plan—pronouns are important. The second, newer function of pronouns is to convey a person's inner sense of themselves as gendered. Here choosing a pronoun is part of one's evolving personhood and their curated gender expression. As I write, neither *she* nor *he* is right to Hiltz—like a given name might feel wrong—and they've asked that we respect that. Easily done, I say.

The way I think about it is that once we have the information we need, it's usually best to be supportive of the other person. We shouldn't be made to ignore our interests in having information about sex, given that it's adaptive in multiple ways. But once we've dealt with those, unless there are other reasons that might make it personally costly for us to go along, we should be kind (because trans people feel cognitive dissonance when they hear what is for them the wrong pronoun) and respectful (because that's what makes a community work).

You may not agree with me on this, either because you think that trans people's interests should always come first or because you think they should always be ignored. But there's one other reasonable concern. Beyond the individuals who are simply asking for kindness and respect, there are those who are using pronouns to advance their sex-blind agenda. If you've read these last few paragraphs and been surprised to remember that pronouns serve an information function unrelated to trans people, you get a sense of the power of their campaign. As I was writing this, I googled "functions of gendered pronouns" and at least on the first page of results, only their second, trans-related function came up. It was as though, before the trans advocacy groups began to push for preferred pronouns as expressions of gender identity, they had no raison d'être. *When people disagree with the advocacy groups on this it's almost never because they don't care about trans people; it's almost always because they also care about sex.*

I know I'm taking a chance, but always holding two thoughts at once and carefully, I'll continue to use people's preferred pronouns when I know who they are. I'll keep saying "no" to formal pronoun campaigns that have the replacement of sex with gender as their goal. And I'll always be looking for other ways to signal that I'm an ally to trans people themselves. If being a good ally—like being a good girl—requires me to be sex-blind, that's not possible. But I will still see you.

Being a Woman Is about More than Gender Identity

This last entry brings me full circle to the beginning of the book. It's my response to the question "What is a woman?" and how I got here. I'm pretty sure that some of this will change, though, because the ask from trans people and their advocates is itself in flux. It's also likely to change because being a woman has a cultural component—as we saw from Myra Bradwell's story in chapter 2, it's often been about more than just the biology—and culture is always changing.

Because sex matters for procreation, sexuality, childrearing, social groups, beauty, aging, disease, medicine, discrimination, and beyond, we need a word for XX people. Even if you disagree that sex is binary—because you're focused on the five hundred reported cases of people with ovotesticular disorder or you've chosen to dissociate sex from reproduction—reproduction itself is undoubtedly binary and so we still need a word for people whose bodies develop to gestate, breastfeed, and care for babies. Because it's comprehensive, efficient, and widely embraced, it makes sense to keep using the word *female* for this purpose.

This definition isn't designed to exclude trans women or to include trans men—that is, to be disrespectful to people in those two groups and their sense of themselves. As it was before transgender issues became part of our lives, it's simply to distinguish XY people from XX people. Saying "male" or "female" codes for a bunch of relevant stuff about a person's phenotype, biology, and experience. The information isn't perfect in individual cases, but it's high-quality.

The word *woman* has been used synonymously with *female* for centuries. We continue to rely on it to signify female. It makes sense that its definition would continue to include people who are female. Outside of certain academic and advocacy circles, the "construct" that is *woman* is undoubtedly built on a physical foundation, and that foundation is female.

As with *female*, this definition of *woman* predates any preoccupation with trans people. Female biology is simply the way we've always anchored the word as it exists to convey certain essential information. ChatGPT's move to disconnect *woman* from *female*—and to ground the definition only in gender identity—was dumb for many reasons, but mainly because for most people, gender identity isn't central or even relevant. Someone can usefully look at me and say, "That woman over there," without having or needing any information about my inner sense of myself as gendered. This should be easy.

The question whether a trans woman is a woman used to be easy too, or at least it was for me, because "trans woman" used to mean an XY person who took female gender-affirming hormones or had feminizing surgery. *Transition* meant physical transition, and this connected trans women to how we understand the word *woman*. We might continue to distinguish trans women from females for the reasons sex will always be important, but the inclusion of physically transitioned trans women in the category "women" makes sense because there's still an important hook to female biology.

To me this is clearly true with respect to XY people who are passing as female. As we saw in chapter 1, this can happen when an XY person is insensitive to their androgens—they have complete androgen insensitivity syndrome—so their body doesn't full develop according to the male design. They are male but we wouldn't call them a man because that's simply not what we see. It also happens when a person's natural form together with comprehensive gender-affirming medical treatment—including surgery—builds a phenotypically female face and body. Again, it doesn't make sense to describe the person we see as a man. When we know it and we do, it's just mean.

A note about the term *passing*. In the United States, it's usually used in connection with very light-skinned black people who are taken for white. Passing was a big deal through the centuries of slavery and the Jim Crow era because of the legal status differences and strict sorting between white and black people. You'll still hear the term used in places where the divisions remain socially significant. At a family funeral in New Orleans several years ago, someone asked about the child of a former neighbor and the response was, "She's passed over, so we don't see her anymore." "Passed over" as in "to the white side." Nella Larsen's heart-wrenching novel *Passing* (1929)—a part of the Harlem Renaissance canon—is helpful, along with Ilyon Woo's *Master Slave Husband Wife* (2023), to understand the historical versions of this phenomenon.

Trans advocacy groups don't like us to use the term *passing* in connection with trans people because they deny any intent to deceive. Although the first clause usually goes unsaid, their point seems to be that unlike a black person passing as white, a trans woman isn't passing as a woman; she is a woman. As we saw already in chapter 3, the movement's relationship with race can be tortured sometimes.

Passing can be about deception but usually it's not; usually we're just going about our lives and others classify us for their own purposes as they go about theirs. We may not even know we're passing. America's "one-drop rule"—one black ancestor colors your blood—together with the tradition in white families of erasing those ancestors, means that millions of white people in this country are passing without knowing it. People who use Ancestry.com to see if they're connected to Thomas Jefferson sometimes find out that they are, but not always to the side of the family they expect. The fact is that human taxonomies are ubiquitous, and people sometimes care to know when someone "passes over" them because they provide useful—workaday or even celebratory—information beyond what the individual needs for themselves.

There are, of course, trans women who have transitioned physically but aren't passing. We generally see them as trans women. We can continue to call them that, but in my experience, so long as we can distinguish when we need to, we usually drop the adjective and that's fine. If you push yourself to understand why you do that, my guess is that you'll realize it's some combination of having the information you need, believing they're a woman because it makes sense to you that gender identity is considered, or knowing that the other person's life will be easier—there will be more joy—when you do.

The question got hard when the definition of a trans woman became—as Cambridge Dictionaries finally put it in print in 2022—"an adult who lives and identifies as female though they may have been said to have a different sex at birth." I discussed Cambridge's questionable relationship to reality and sex in chapter 3, but what's notable here is that the definition severed the connection between being trans and transitioning physically. It made being trans about social transition—"coming out as trans"—instead. This was a well-intentioned move, designed to account for trans people who couldn't or didn't want to "medicalize" their transness. In the process, however, it also severed the connection between woman and the female form and experience.

Trans people aren't of one voice about the acceptability of distinguishing among them based on whether they've had gender-affirming medical treatment. Those who think distinguishing isn't acceptable are more visible. They've coined the derogatory term *transmedicalist* to mean someone who says that being trans is having and physically treating gender dysphoria. To many trans people, transmedicalism is anti-trans hate whether it's coming from other trans people or from outsiders like me. To them, this whole discussion is transphobic because it recognizes physical differences among trans women.

I'm writing nonetheless because for the rest of society—especially for females who aren't trans—severing the connection between woman and the female form and experience is a monumental conceptual move. *This needs to be said.* Over the centuries-long history of the Woman Question, our predecessors worked to rid *woman* of the artifice—*including the idea of an inner feminine self*—that attached to her because of her sex. They weren't working to rid *woman* of sex itself.

It's also a monumental practical move. In the qualifier "lives as female" we might expect to see feminine gender expression and with it some female experience. But because many of us—including trans people—eschew gender conformity in this period, we might not. In that case, we could no longer expect to get any of the information we depend on.

When people describe this change in the definition of a trans woman as an "erasure" it's because it is. The question for those who also care about sex is whether it's too much of one.

My friend Joanna Harper is helpful here. Joanna is a scientist who works on sex differences in human performance. She focuses on the nature and extent of the "retained" or "legacy" advantages of male puberty after trans women—like Lia Thomas—take female gender-affirming hormones.

Joanna is also a competitive athlete who was born male. She remembers from when she was a little girl always having an inner sense of herself as female, but she transitioned only later because transition meds weren't available when she was going through puberty, because her parents wouldn't have been supportive anyway, and because—like Caitlyn Jenner before her—she was trying to make it work. Life is hard and Joanna's has been unbelievably so, most recently because she's been attacked and undermined by people who can't abide her fact-based work. But about being a woman, she's sure and shining. She explains, "Trans women are a subcategory of women. We're not physiologically the same as cis women [females] but we are women." Then, about trans women who have only just come out socially, she adds, "You have to start somewhere." She sounds a lot like Cate Campbell, who says, "Trans women are women but they're not female."

I'm usually comfortable with this resolution. "Trans women are women"—without the full stop—can work for almost everyone. But then I go back to struggling, first with the equation of two things that aren't at all alike, and then with the broader implications of divorcing *woman* from the female physical experience, and I throw in the towel. Sometimes I can't get past the nuance, and it's a good thing there are other smart, caring people on the case.

The Woman Question became easy again when progressive advocates abandoned the qualifier "lives as a woman" and insisted that "identifies as a woman" is all that it takes to be one. No female biology, phenotype, or experience necessary. No requirement that you present as a woman. It's only and all about your inner sense of yourself. Beyond that, those of us on the outside are prohibited from communicating anything about a person's body. We can't say "no" even to repeat woman killer Harvey Marcelin, aka Marceline Harvey, who presents as a man in a women's shelter. Doing so is anti-trans hate and that charge ends any discussion.

But it's not anti-trans to reject this overreach; it's human. It's also why females as far apart from each other as Marsha Blackburn and

Kathleen Stock put themselves on the line to call it out. Arguably the only things a conservative, heterosexual, country club Tennessean and a liberal, lesbian, academic Scot have in common are their sex and this hard NO. They're also both white, but don't be misled. This isn't about race. Among many others, there's the award-winning feminist writer Chimamanda Ngozi Adichie, who was for a time engaged in a public quarrel, mostly with trans activists in the United Kingdom, because she said that "trans women are trans women."

Because of what *woman* already means, however nice you want to be, this last erasure—"an adult who identifies as female" rather than "*lives and* identifies"—is category defeating. A word that's designed to communicate information, say about the color red, can't be defined as red and as blue. Outside of certain privileged circles, it's nonsense—and offensive and painful and frightening—to insist that an unadulterated male who presents in a masculine way is a woman. When people say otherwise, as Fatima Mernissi put it about colors in the quotation I used as an epigraph, it's clear we're no longer talking about the same thing.

Mary Wollstonecraft is famous for being both the mother of modern feminism with her *A Vindication of the Rights of Woman* (1792), and the mother of Mary Shelley, author of the Gothic masterpiece *Frankenstein* (1818). Wollstonecraft died in 1797, at the age of thirty-eight, just eleven days after Mary was born. The cause was "childbed fever," the result of postpartum hemorrhage followed by an infection in a bit of retained placenta.

Vindication is a leg in the centuries-long relay that is the Woman Question. You'll recognize its themes as those with which I began this book: "[E]ither Nature has made a great difference between man and man, or that the civilization which has hitherto taken place in the world has been very partial," Wollstonecraft wrote. She recognized as natural (and celebrated) man's relative physical strength:

In the government of the physical world it is observable that the female in point of strength is, in general, inferior to the male. This is the law of Nature; and it does not appear to be suspended or abrogated in favor of woman. A degree of physical superiority cannot, therefore, be denied, and it is a noble prerogative!

She also recognized as natural (and honored) women's reproductive role. But she insisted that it doesn't justify their different "civilization"—meaning their different upbringing, education, and opportunities—much less the different expectations we have of their characters:

Women, I allow, may have different duties to fulfil; but they are human duties, and the principles that should regulate the discharge of them, I sturdily maintain, should be the same.

Throughout *Vindication*, Wollstonecraft argued that if women of the higher classes—a distinction she had to draw in the eighteenth century—lacked reason, virtue, and experience, unnatural obstacles were to blame, not any inherent incapacity. She was all about ridding our notion of *woman* of such unnatural obstacles.

Following her death in 1797, Wollstonecraft was publicly ridiculed by men and women alike for her unorthodox "unsexed" lifestyle—which included lovers and a first child out of wedlock—and for her writings *against* the patriarchy and eighteenth-century notions of womanhood, and *for* the coeducation of girls and boys. The effect of their ridicule was as intended: her work was condemned and then covered up by the conventions of the day. She was read nonetheless, even if on the sly, as *Vindication* was published throughout the nineteenth century, and we see its unbroken influence from Jane Austen through the suffragists to modern-day feminism. The women's movements of the twentieth century fully resurrected her, although the public legacy of her original condemnation has been durable: she's remarkably undercelebrated in relation to her contributions.

In 2010, a group called "Mary on the Green" set out to remedy that in an area of North London where she had lived and worked. A decade later, in 2020, in a park in Newington Green, they unveiled A *Sculpture for Mary Wollstonecraft*. As the artist Maggi Hambling intended it, the work represents Wollstonecraft's "ideals" and "what she made happen." For Hambling, Wollstonecraft sparks a "vital contemporary discourse for all that is still to be achieved" in relation to women's rights.

Mark Brown of the *Guardian* described Hambling's sculpture as "a silvery naked everywoman figure emerging free and defiantly from a swirling mingle of female forms." To me, it conjures still-sexually indistinct embryonic development or unformed physical matter from which woman emerges in her utterly natural state, the one that's completely denuded of gender. The work isn't purely beautiful but, like the *David* and the female form in Robert Graham's *Olympic Gateway*, the suggestion is that she is fit and strong.

The fact that she's fit and nude made the sculpture "one of the most controversial, most debated and most polarizing public artworks of 2020," Brown wrote: "Social media went bananas. What the hell was it? Why was the mother of feminism being celebrated with a naked Barbie doll at the top of it? Surely a colossal error of judgment?" One passerby found "the nakedness jarring. I don't understand why it was necessary for [Hambling] to include a naked female form for a feminist icon." Another said, "I don't really get the lady at the top if it's all about women and empowering women. . . . I don't get why she's naked."

The thing is that the business of covering our bodies doesn't come from Wollstonecraft. Its source is a combination of traditional modesty norms (which she eschewed) and later strands of feminism (which she couldn't know) that similarly require it: To get men, and women and girls who are interested in men's attention, to focus elsewhere. To discourage body shaming. Or to disconnect women from sex altogether. The theory is that what's covered doesn't exist or can't hurt us. Caroline Criado Perez has shown us that nothing could be further from the truth, but in the meantime, they've managed again to cover Wollstonecraft—by projecting their conventions onto her text and image.

I encourage you to read *Vindication* for yourself. I think you'll find that the power and the beauty of a "strong" not "libertine" female was enduring for her too. She distinguishes strength from weakness and emphasizes that she means both in body and mind. She speaks directly to the cultural and physical "confinement" of females relative to males, and of how "the body" confined "is prevented from attaining that grace and beauty which relaxed half-formed limbs never exhibit." And she concludes with this eternal point on sex and gender:

> [A] girl, whose spirits have not been dampened by inactivity, or in-nocence tainted by false shame, will always be a romp, and the doll will never excite attention unless confinement allows her no alterna-tive. Girls and boys, in short, would play harmlessly together, if the distinction of sex was not inculcated long before nature makes any difference. I will go further, and affirm, as an indisputable fact, that most of the women, in the circle of my observation, who have acted like rational creatures, or shown any vigor of intellect, have acciden-tally been allowed to run wild, as some of the elegant formers of the fair sex would insinuate.

I can relate. Sex is fixed, but as gender goes, *il y a plusieures manières de construire un être humain.* Beyond this, if we retain any-thing from Wollstonecraft and the women who held and passed the baton across the centuries, it's that it's not feminist to tell a female to stand down, and when she doesn't, to tell people not to listen to what she has to say.

Acknowledgments

It literally takes a village to make a book, and this one, which involves science, law, politics, art, sports, and so much more, and which seeks to forge a path through deeply partisan divides, took an especially large one. Everyone who helped me in one way or another, whether they're named or not, has my sincere gratitude.

This begins with my family. To my sons, Alexander and Nicolas, for their generational voice, for reading drafts, and for research assistance, thank you. We joke, sort of, that my husband, Jim, let me put his family name on the cover despite the personal narrative and the target it puts on his back too; but like the great defense lawyer he is, he's always—in his words—"a happy warrior." I can't overstate how valuable it is to be surrounded by good, strong, feminist men. There are two books—one fiction, one nonfiction called *The End of Men*. If ever a case needed to be made against that bad idea, mine are it.

Beyond family, I can't overstate how valuable it's been to work with the best experts in their respective domains. This starts with my editor, Robert Messenger. Robert is incomparable at his craft but could easily have a second career coaching—his own boys are lucky for that. There's also my magic agent, Jim Levine, who got me to believe the impossible—that this project was viable in the current environment—and then made it happen. "Trust me," he said, right before he delivered.

I am indebted to the experts—scientists, lawyers, scholars, service members, and others—who took the time to educate me whether it was on or off the record. Among them were Art Arnold, Mike Brown, Chandler Cole, Joanna Crisman, Jolynn Dellinger, Tom Ferraro, Raffi Grinberg, Denise Harle, Angelica Hirschberg, Sandra Hunter, Michael Joyner, Ben Levine, Virginia Miller, Ted Olson, Jeff Powell, Wickliffe Shreve, Neil Siegel, Quinn Ostrom, David Toole, and Kyle Walsh. There were many others; I trust that they know who they are. Altogether, their help speaks to their collegiality not to their agreement with me. Any errors in the presentation are mine.

No less important were the professionals who produced the book and the staff who supported me as I was writing. At Duke Law School, Tiffany Kollock kept my computers working and managed my technology crises with invaluable skill and steadiness; and Michelle Shields, Marlyn Dail, and Susan Ranes made sure that I had what I needed administratively as I juggled the combination of teaching and writing. At Simon & Schuster, Jackie Seow nailed the perfect cover; Thomas Pitoniak returned an exceptional copyedit; Ruth Lee-Mui gave the book a gorgeous interior design; Lisa Healy managed the production with unfailing aplomb; Elizabeth Herman and Stephen Bedford were all I could have hoped for in a publicist and marketer; and Isabel Casares was terrific as mission control as the entire team made the manuscript into a book. I can't thank you all enough.

I am also grateful to extended family, friends, and colleagues for being there in various essential ways. Nita Farahany convinced me to lean into the project when it would have been easier to walk away and then generously shared her connections and experience. Lisa Griffin, Liz Gustafson, and Anne Phelps made it possible for me actually to do it. David Boyd, Lalou Dammond, Helena Guenther, and Kathleen Stock read drafts and provided feedback that was—in no small part because of their different perspectives—indispensable. Donna de Varona, Nancy Hogshead, Donna Lopiano, Martina Navratilova, and Tracy Sundlun included me on their team for a time; I learned so much from them and am in awe of their decades-long devotion to girls' and women's sports.

Beyond writing, the core of my work is teaching, and anyone who does this knows that the learning always goes both ways. Over the past few years, my students at Duke Law have taught me by the combination of their trust and their resistance that passion for ideas, justice, and intellectual engagement lives in their generation. A special shout-out to those who participated in the Culture Wars and Illiberalism seminars who shared their knowledge, experience, and perspectives, and who demonstrated every day that political partisans can still sit down, break bread, and talk openly, civilly, and productively about the most difficult of topics. My gratitude also to the women who came to the house to talk about how their generation thinks about having kids and motherhood; and to the men who shared their commitment not just to modeling but also to creating spaces for conversations about modern masculinity. Their thoughtfulness abounds.

As an author of a book that touches on The Woman Question, I stand on the shoulders of many others who, over the centuries but especially in the last half century, have written on it too, in one way or another. I was only able to highlight a tiny fraction. This means that when I look at my bookshelves, I see names like Natalie Angier, Susan Faludi, Germaine Greer, Joan Kelly—the list goes on and on—and I feel remiss. For historical perspective, including so that we don't continue to make the mistake of thinking that a handful of women in the European diaspora invented feminism in the late nineteenth century, and for the timelessness of the wisdom and arguments, we need more opportunities to tell the story of how this modern work sits in its whole frame. I hope that my nod to Morocco's Fatema Mernissi suggests one way into that necessary adventure.

Finally, to Kerry Abrams, my dean at Duke Law, and to Jonathan Karp and Priscilla Painton at Simon & Schuster. I've never spoken with them about whether they agree or disagree with my positions, but I know they believe in the importance of the respectful exchange of ideas and about doing and supporting work that can contribute to mutual understanding and healing our partisan divides. Whatever you think of this book, we all benefit from their community-minded leadership.

Bibliography

Abramson, Paul, and Steven Pinkerton. *With Pleasure: Thoughts on the Nature of Human Sexuality*. New York: Oxford University Press, 1995.

Brizendine, Louann, M.D. *The Female Brain*. New York: Broadway Books, 2006.

———. *The Male Brain*. New York: Three Rivers Press, 2010.

Clark, Kenneth. *The Nude: A Study in Ideal Form*. New York: Doubleday, 1956.

Clarke, Edward H. *Sex in Education, or A Fair Chance for Girls*. Boston: Houghton, Mifflin, 1873.

Collins, Robert. *Essay on the Treatment and Management of Slaves*. 2nd ed. Boston: Eastburn's Press, 1853.

Criado Perez, Caroline. *Invisible Women: Data Bias in a World Designed for Men*. New York: Abrams Press, 2019.

Doniger, Wendy, and Sundhir Kakar. *Kamasutra*. Oxford: Oxford University Press, 2002.

Friedman, Jane M. *America's First Woman Lawyer: The Biography of Myra Bradwell*. Buffalo, NY: Prometheus Books, 1993.

Hooven, Carole. *Testosterone: The Story of the Hormone That Dominates and Divides Us*. New York: Henry Holt, 2021.

Institute of Medicine, Committee on Understanding the Biology of Sex and Gender Differences. *Exploring the Biological Contributions to Human Health: Does Sex Matter?* Washington, DC: National Academy Press, 2001.

Kelly, Joan. *Women, History, and Theory: The Essays of Joan Kelly*. Chicago: University of Chicago Press, 1984.

Mernissi, Fatima. *Dreams of Trespass: Tales of a Harem Girlhood*. New York: Basic Books, 1994.

Painter, Nell Irvin. *Sojourner Truth: a Life, a Symbol*. New York: Norton, 1996.

Paoletti, John. *Michelangelo's David: Florentine History and Civic Identity*. Cambridge: Cambridge University Press, 2015.

Reeves, Richard V. *Of Boys and Men: Why the Modern Male Is Struggling, Why It Matters, and What to Do about It*. Washington, DC: Brookings Institution Press, 2022.

Stock, Kathleen. *Material Girls: Why Reality Matters for Feminism*. London: Fleet, 2021.

Wollstonecraft, Mary. *A Vindication of the Rights of Woman*. London: Penguin Books, 1988.

Notes

Introduction

xix   *"just fears and myths and stereotypes"*: H.R. 5, The Equality Act, Hearing before the Committee on the Judiciary, House of Representatives, 116th Congress, 1st Session, April 2, 2019, Serial No. 116-13, Testimony of Sunu Chandy, Transcript, 96 and 143.

xix   *Here's the quote*: H.R. 5 Hearing Transcript, 156.

xx   *Megan Rapinoe*: Ben Morse, "Megan Rapinoe Says US Has 'Weaponized' Women's Sports against Trans People, 'Trying to Legislate Away People's Full Humanity,'" CNN, July 10, 2023, https://www.cnn.com/2023/07/10/sport/megan-rapinoe-trans-rights-us-soccer-spt-intl/index.html.

xx   *Caitlyn Jenner*: Glenn Garner, "Caitlyn Jenner Faces Backlash for Opposing Trans Girls in Sports: 'I'm Clear About Where I Stand,'" *People*, May 2, 2021, https://people.com/sports/caitlyn-jenner-opposing-trans-girls-in-sports-clear-about-where-i-stand/.

1: The Answer from Biology

3   *sport scientist Ross Tucker*: Ross Tucker, "The Caster Semenya Debate," *Science of Sport*, July 16, 2016, https://sportsscientists.com/2016/07/caster-semenya-debate/.

3   *a headline afterward*: Cathal Dennehy, "Caster Semenya Interview: 'I Am the Greatest That Has Ever Done It,'" *Irish Examiner*, July 4, 2022, https://www.irishexaminer.com/sport/othersport/arid-40910609.html.

4   *I was in the race*: Jeré Longman, "Track's Most Resilient (and Suspect) Record Is in Danger," *New York Times*, June 15, 2017, https://www.nytimes.com/2017/06/15/sports/olympics/jarmila-kratochvilova-800-meters-record.html.

4   *Semenya has come to represent*: Melissa Block, "'I Am a Woman': Track Star

Caster Semenya Continues Her Fight to Compete as Female," NPR, May 31, 2019, https://www.npr.org/2019/05/31/728400819/i-am-a-woman-track-star -caster-semenya-continues-her-fight-to-compete-as-a-female.

5   *"Semenya's grandmother, Maphuti Sekgala"*: "Caster Semenya's Mother Hits Out at Gender Dispute," *Guardian*, August 20, 2009, https://www.theguardian.com /sport/2009/aug/20/caster-semenya-gender-world-championship-dispute.

5   *He recalls stopping to use the facilities*: "Caster Semenya's Mother."

5   *"man-like physical features"*: See, for example, Gardy Chacha, "Margaret Nyairera and the Gender Agenda," *Standard*, August 31, 2016, https://www.standardme dia.co.ke/entertainment/sunday-magazine/article/200.

5   *As Semenya herself tells the story*: "Caster's Love Story with Her Wife: She Thought I Was a Boy," *Sunday Times*, August 8, 2017, https://www.timeslive .co.za/tshisa-live/tshisa-live/2017-08-08-casters-love-story-with-her-wife-she -thought-i-was-a-boy/.

5   *The video of her victory*: "Semenya Becomes World Champion—from Universal Sports," YouTube, accessed September 14, 2023, https://www.youtube.com /watch?v=MpblUehi9Dk.

5   *her primary school trainer*: Ariel Levy, "Either/Or," *New Yorker*, November 19, 2009, https://www.newyorker.com/magazine/2009/11/30/eitheror.

6   *The headmaster at her high school*: "Caster Semenya's Mother."

6   *When Semenya crossed the finish line*: Shane Aaron Miller, "'Just Look at Her!': Sporting Bodies as Athletic Resistance and the Limits of Sport Norms in the Case of Caster Semenya," *Men and Masculinities* 18, no. 3 (2015): 271–92.

6   *South African sportswriter Wesley Botton*: Donald McRae, "The Return of Caster Semenya: Olympic Favourite and Ticking Timebomb," *Guardian*, July 29, 2016, https://www.theguardian.com/sport/2016/jul/29/the-return-of-caster-seme nya-olympic-favourite-and-ticking-timebomb.

8   *The WHO, for example, defines gender*: "Gender," World Health Organization, https://www.who.int/europe/health-topics/gender.

8   *The IOM explains*: Institute of Medicine, *Exploring the Biological Contributions to Human Health: Does Sex Matter?* (Washington, DC: National Academies Press, 2001), 1, fn 1.

9   *Human dimorphism is less pronounced*: Carole Hooven, *Testosterone: The Story of the Hormone That Dominates and Divides Us* (New York: Henry Holt, 2021), 145.

12 *in an XX embryo*: "What Is a Vaginal Self-Exam," WebMD, March 2023, https:// www.webmd.com/women/what-is-a-vaginal-self-exam.

15 *pubertal developments that both males and females*: Max Roser, Cameron Appel, and Hannah Ritchie, "Human Height," Our World in Data, 2013, https:// ourworldindata.org/human-height.

16 *Males' energy metabolism favors muscle*: "The Role of Testosterone in

Athletic Performance," January 2019, available at https://law.duke.edu/sites /default/files/centers/sportslaw/Experts_T_Statement_2019.pdf.

16 *mechanism for the physical changes that occur in males:* Jonathon W. Senefeld et al., "Divergence in Timing and Magnitude of Testosterone Levels Between Male and Female Youths," *Journal of the American Medical Association* 321, no. 1 (July 7, 2020): 99–101. There is a false controversy over the question whether male and female testosterone levels overlap. The data are clear that absent a rare disorder, starting from the onset of male puberty, they do not. Notably, although females with polycystic ovaries have higher testosterone levels than females who don't have this condition, their levels are still far below the bottom of the normal male range. This controversy—and the way it's been generated—is discussed further at the close of chapter 1.

18 *It's also why transgender boys:* Adams v. School Board of St. Johns County, 18-13592, 304-1, (11th Cir. 2022).

19 *Our brains are notably plastic:* Louann Brizendine, *The Male Brain* (New York: Harmony/Rodale, 2010), 5–6.

19 *evolutionary biologist Carole Hooven:* Hooven, *Testosterone*, 167.

20 *It shouldn't need to be said:* Louann Brizendine, *The Female Brain* (New York City: Mordan Road Books, 2006), 6–8.

21 *nature remains a powerful force:* Brizendine, *The Female Brain*, 6–7.

22 *Rather, they experience a gradual:* "Male Menopause: Myth or Reality?" Mayo Clinic, May 24, 2022, https://www.mayoclinic.org/healthy-lifestyle/mens -health/in-depth/male-menopause/art-20048056.

23 *When the deviations cause functional impairments:* "Why Is ISNA Using DSD?" Intersex Society of North America, accessed September 14, 2023, https://isna .org/node/1066/.

25 *Intersex Human Rights Australia:* Admin, "Intersex Population Figures," Intersex Human Rights Australia, last modified September 16, 2019, https://ihra.org .au/16601/intersex-numbers/.

25 *atypical brain development:* See, for example, Joshua D. Safer, "A Current Model of Sex Including All Biological Components of Sexual Reproduction," *Law and Contemporary Problems* 85, no. 1 (2022): 47–56.

26 *Within milliseconds we perceive:* See, for example, Frank E. Pollick et al., "Gender Recognition from Point-Light Walkers," *Journal of Experimental Psychology: Human Perception and Performance* (December 2005):1247–65.

27 *the whole topography of the face:* Alessandro Cellerino, Davide Borghetti, and Ferdinando Sartucci, "Sex Differences in Face Gender Recognition in Humans," *Brain Research Bulletin* 63, no. 6 (July 2004): 443–49; Christian Kaul, Geraint Rees, and Alumit Ishai, "The Gender of Face Stimuli Is Represented in Multiple Regions in the Human Brain," *Frontiers in Human Neuroscience* 4 (2010).

27 *Stanford neurobiologist Nirao Shah*: Bruce Goldman, "Animal Magnetism," *Stanford Medicine Magazine*, June 17, 2019, https://neuroscience.stanford.edu /news/animal-magnetism.

27 *For years, Semenya's allies*: Marcie Bianco, "Olympic Champion Caster Semenya's Critics Couch Misogynoir in the Language of 'Equality,'" NBC News, September 10, 2020, https://www.nbcnews.com/think/opinion/olympic-champion -caster-semenya-s-critics-couch-misogynoir-language-equality-ncna1239780.

28 *One writer argued, for example*: Tessa Stapp, "Double Standards in the Olympics and Beyond: Marginalized Groups Are Running an Entirely Different Race," *Berkeley Political Review*, January 4, 2021, https://bpr.berkeley.edu /2021/01/04/double-standards-in-the-olympics-and-beyond/.

28 *Another writer warned*: Bianco, "Olympic Champion."

29 *Going into the Rio de Janeiro Olympics*: "Who Gets to Race as a Woman?" editorial, *New York Times*, August 19, 2016, https://www.nytimes.com/2016/08/20 /opinion/who-gets-to-race-as-a-woman.html.

31 *one of the most brilliant judicial passages*: CAS 2018/0/5794 Mokgadi Caster Semenya v. International Association of Athletics Federations, and CAS 2018/0/5798 Athletics South Africa v. International Association of Athletics Federations, (Court of Arbitration for Sport 2019), https://www.tas-cas.org/file admin/user_upload/CAS_Award_-_redacted_-_Semenya_ASA_IAAF.pdf.

32 *To her longtime chronicler*: Lynsey Chutel and Jeré Longman, "The Clock Ticks on Caster Semenya's Olympic Career," *New York Times*, June 28, 2021, updated August 7, 2021, https://www.nytimes.com/2021/06/28/sports/olympics/cas ter-semenya-olympics-gender.html.

32 *Unapologetically herself*: Karen Zraick, "Caster Semenya, Hero in South Africa, Fights Hormone Testing on a Global Stage," *New York Times*, May 1, 2019, https://www.nytimes.com/2019/05/01/sports/who-is-caster-semenya.html.

2: The Answer from Law

35 *In the heat of battle*: Jane M. Friedman, *America's First Woman Lawyer: The Biography of Myra Bradwell* (New York: Prometheus, 1993), 131, 139.

35 *Myra Bradwell was having none of it*: Friedman, *Myra Bradwell*, 129.

35 *She had been raised by parents*: Friedman, *Myra Bradwell*, 35.

36 *Lavinia Goodell of Wisconsin*: Friedman, *Myra Bradwell*, 149.

36 *her petition in Bradwell v. State of Illinois*: Friedman, *Myra Bradwell*, 24.

36 *as she put it in the Chicago Legal News*: Friedman, *Myra Bradwell*, 25.

36 *claims about the nature of males*: Friedman, *Myra Bradwell*, 23.

37 *the Nation declared Bradwell's arguments*: Friedman, *Myra Bradwell*, 27.

37 *The Supreme Court itself chose to avoid sex entirely*: Bradwell v. Illinois, 83 U.S. 130, 133 (1872).

37 *One lone justice chose not to skirt the question*: Bradwell v. Illinois, at 142.

39 *"the police power"*: The table of contents of the California code is illustrative of the scope of the government's police powers. It's available at https://law.justia .com/codes/california/2021/.

42 *social encounters are typically "predicated"*: Goldman, "Animal Magnetism."

43 Black's Law Dictionary: The current edition is Brian A. Garner, *Black's Law Dictionary*, 11th ed. (St. Paul, MN: Thomson West, 2019). The first, second, and third editions (also by Henry Campbell Black) were published in 1891, 1910, and 1933 respectively.

44 Bostock v. Clayton County: Bostock v. Clayton County, 590 U.S. ___ , 140 S. Ct. 1731 (2019).

44 *Following* Bostock, *a federal appellate*: West v. Radke, No. 20-1570 (7th Cir. 2022).

45 *often credited to Ruth Bader Ginsburg*: "Sincerely, Ruth Bader Ginsburg," Columbia Law School, 2018, accessed September 14, 2023, https://www.law.colum bia.edu/sites/default/files/2020-06/20180921_sincerely_rbg.pdf.

45 woman *and* man *as synonyms for* female *and* male: Kathleen Stock, *Material Girls: Why Reality Matters for Feminism* (London: Fleet, 2021).

46 *The California Health and Safety Code*: 2007 California Health and Safety Code Art. 2, Sec. 102425-102475, Content of Certificate of Live Birth, https://law.justia.com/codes/california/2007/hsc/102425-102475.html#: ~:text=102425.,day%2C%20hour%2C%20and%20year.

46 *the Family Division of the High Court*: The Queen (on the Application of TT) and the Registrar General for England and Wales, Case No: FD18F00035 (2019), EWHC 2384 (Fam), https://www.judiciary.uk/wp-content/uploads /2019/09/TT-and-YY-APPROVED-Substantive-Judgment-McF-23.9.19 .pdf.

48 *It was and remains fine, for example, to infer*: Office on Women's Health, "Premenstrual Syndrome," U.S. Department of Health and Human Services, accessed September 14, 2023, https://www.womenshealth.gov/menstrual-cycle /premenstrual-syndrome.

48 *By contrast, it was dumb to leap*: Joan Kelly, *Women, History, and Theory: The Essays of Joan Kelly* (Chicago: University of Chicago Press, 1984), 65–95.

48 *Dumb too were the pseudoscientific elaborations*: Friedman, *Myra Bradwell*, 37.

49 *Lest you think this some outlier*: Edward H. Clarke, *Sex in Education, or A Fair Chance for Girls* (Boston: Houghton, Mifflin, 1873).

50 *Whether males can hold hands*: Lana Berkowitz, "Bush, Prince Showed Respect by Holding Hands," *Houston Chronicle*, April 27, 2005, https://www.chron.com /life/article/Bush-prince-showed-respect-by-holding-hands-1948636.php.

50 *Whether males cry at all or in public*: Peter Bregman, "Nadal Is Strong Enough

to Cry. Are You?" *Harvard Business Review*, September 11, 2013, https://hbr
.org/2013/09/nadal-is-strong-enough-to-cry.

50 India Times *was on the same case:* Basit Aijaz, "'Real Men Get Emotional': Rafael Nadal Crying for Roger Federer Leaves Internet Heartbroken," *India Times*,
September 24, 2022, https://www.indiatimes.com/trending/social-relevance
/roger-federer-rafael-nadal-crying-laver-cup-internet-emotional-580455.html.

51 the *"average enslaved woman"*: Jennifer Hallam, "Slavery and the Making of America: The Slave Experience: Men, Women, & Gender," PBS, 2004, https://
www.thirteen.org/wnet/slavery/experience/gender/history.html.

51 the *"typical American woman"* living in freedom: S. Mintz and S. McNeil, "Limiting
Births in the Early Republic," Digital History, 2021, https://www.digitalhis
tory.uh.edu/topic_display.cfm?tcid=134.

52 she told the *Above the Law blog*: Aris Folley, "Woman Who Went Viral after
Taking Bar Exam While in Labor Finds Out She Passed," *Hill*, December 1,
2020, https://thehill.com/blogs/blog-briefing-room/news/528269-woman-who
-went-viral-after-taking-bar-exam-while-in-labor-finds/.

54 the clerk's list of *"Lady Lawyers"*: "In Re Lady Lawyers: The Rise of Women Attor-
neys and the Supreme Court," Supreme Court of the United States, accessed
September 14, 2023, https://www.supremecourt.gov/visiting/exhibitions/La
dyLawyers/Default.aspx.

54 *She died two years later:* Friedman, *Myra Bradwell*, 30.

3: The Answer from Progressive Advocacy

56 *Most familiar was ordinary servitude:* Robert Collins, *Essay on the Treatment and
Management of Slaves*, 2nd ed. (Boston: Eastburn's Press, 1853).

56 *My ancestor Ellen Craft:* My Dad was Henry (Hank) Craft Dammond, Ellen
Craft's great-great grandson through their eldest son Charles. Dad was not my
biological father, but he raised me with my mother from the time I was a toddler until he died when I was eleven. Mom and Dad had a common law marriage and so my relationship to him could be called a common law adoption,
but the distinction between "real" and "adoptive" Dad was never part of our
family story. As was the case for my brother Tokya and my sister Lalou, who
were his biological children, he was the only father—and growing up his was the
only paternal lineage—I knew. I've since met my biological father's family, but I
never met him, and we had no legal ties. He was also black, from Louisiana.

57 *Enslaved black women bearing children:* Eugene R. Dattel, "Cotton and the Civil
War," *Mississippi History Now*, July 2008, https://www.mshistorynow.mdah.ms
.gov/issue/cotton-and-the-civil-war.

57 *Jackson appeared before the Senate Judiciary Committee:* Nina Totenberg, "Ketanji
Brown Jackson, Biden's Supreme Court Nominee, Has Blazed Trails All Her

Life," NPR, February 14, 2022, https://www.npr.org/2022/02/14/1078086453 /ketanji-brown-jackson-supreme-court-biden; https://www.whitehouse.gov/kbj/.

58 *Senator Booker noted*: Hearing Transcript, "Senate Judiciary Committee Holds Hearing on the Nomination of Ketanji Brown Jackson to be an Associate Justice on the Supreme Court of the United States, Day 1," March 21, 2022, http://congressional .proquest.com/congressional/docview/t39.d4.tr03210122.o05?accountid=10598.

58 *A photograph of the hearings went viral*: Gina Cherelus, "The Story Behind That Photo of Ketanji Brown Jackson and Her Daughter," *New York Times*, March 24, 2022, https://www.nytimes.com/2022/03/24/style/ketanji-brown-jackson-da ughter-photo.html.

60 *this fictional set of opposites*: While this line is most often rendered as "Ain't I a woman?" I am choosing to include it here as "Am I not a woman?" because the latter is consistent with a popular abolitionist movement's slogan of the period, with which Truth, at the time a prominent abolitionist herself, would have been familiar. Also, Truth spoke only Dutch as a child and standard English thereafter. To have spoken the phrase differently would have been an affectation. Finally, we know that the more common rendering comes not from Truth's own speech, for which there isn't a transcript, but rather from one given twelve years later by the white abolitionist Frances Dana Parker Gage. Gage's speech, which did make it into the published record, contained different known fabrications about Truth and her life, fabrications that were apparently designed to portray Truth in a way that fit the stereotype of an uneducated slave. For this and more I recommend Nell Irvin Painter's biography, *Sojourner Truth: A Life, A Symbol* (New York: Norton, 1996), 3–20 and 164–78.

61 *the ACLU's Chase Strangio*: Chase Strangio, "Caster Semenya Is Being Forced to Alter Her Body to Make Slower Runners Feel Secure in their Womanhood," NBC News, May 1, 2019, https://www.nbcnews.com/think/opin ion/caster-semenya-being-forced-alter-her-body-make-slower-runners-ncna 1000896.

61 *the National Women's Law Center*: Auden Perino, "NWLC Leads Amicus Brief Challenging Anti-Trans Sports Ban in West Virginia," National Women's Law Center, April 4, 2023, https://nwlc.org/nwlc-leads-amicus-brief-challenging -anti-trans-sports-ban-in-west-virginia/.

64 *changed Ginsburg's word* women *to* people: Michael Powell, "A.C.L.U. Apologizes for Tweet That Altered Quote by Justice Ginsburg," *New York Times*, September 27, 2021, https://www.nytimes.com/2021/09/27/us/aclu-apologizes-gins burg-quote.html.

64 *Michelle Goldberg of the* New York Times: "Michelle Goldberg Grapples with Feminism After Roe," *The Ezra Klein Show*, audio, July 8, 2022, https://pod casts.apple.com/kg/podcast/michelle-goldberg-grapples-with-feminism-after

-roe/id1548604447?i=1000569246636 or https://www.nytimes.com/2022/07/08/opinion/ezra-klein-podcast-michelle-goldberg.html.

68 *"the dead weight of female biology"*: Victoria Smith, "Dirty Feminism: You Can't Have Feminism without Femaleness," Critic, December 13, 2022, https://the critic.co.uk/dirty-feminism/.

69 no *"unfair advantage"*: Human Rights Campaign, "Myths and Facts: Battling Disinformation About Transgender Rights," https://www.hrc.org/resources/myths-and-facts-battling-disinformation-about-transgender-rights.

71 *Jennifer Finney Boylan's 2023 op-ed*: Jennifer Finney Boylan, "To Understand Biological Sex, Look at the Brain, Not the Body," *Washington Post*, May 1, 2023, https://www.washingtonpost.com/opinions/2023/05/01/transgender-biology-brain-science-freedom/.

71 *Judge Jill Pryor's dissent*: Drew Adams v. School Board of St. Johns Co., 18-13592, (11th Cir. 2022), https://www.govinfo.gov/app/details/USCOURTS-ca11-18-13592.

72 *cognitive scientist Lera Boroditsky*: Lera Boroditsky, "How Language Shapes Thought," *Scientific American*, February 1, 2011, https://www.scientificamerican.com/article/how-language-shapes-thought/.

73 *the term* biological sex: Cassius Adair et al., *Style Guide*, Trans Journalists Association, accessed June 16, 2023, https://transjournalists.org/style-guide/; GLAAD, "Glossary of Terms: Transgender," *GLAAD Media Reference Guide*, 11th ed., 2022, https://glaad.org/reference/trans-terms/.

73 *updating the definition of the word* transgender: Department of Health and Human Services, Office of Population Affairs, "Gender-Affirming Care and Young People," accessed September 14, 2023, https://opa.hhs.gov/sites/default/files/2022-03/gender-affirming-care-young-people-march-2022.pdf.

74 *Peyton Thomas did this to Louisa ("Lou") May Alcott*: Peyton Thomas, "Did the Mother of Young Adult Literature Identify as a Man?" *New York Times*, December 24, 2022, https://www.nytimes.com/2022/12/24/opinion/did-the-mother-of-young-adult-literature-identify-as-a-man.html.

75 *the proposed redefinition of sex omits the traditional biological definition*: Oxford English Dictionary, 2nd ed. (Oxford: Oxford University Press, 1989), s.v. "Sex, n.1," https://www.oed.com/viewdictionaryentry/Entry/176989#:~:text=Physical%20contact%20between%20individuals%20involving,in%20sexual%20intercourse%20(with).

76 *the text doesn't mention it*: Following the formula shortage of 2022, Congress separately passed the clearly protective Providing Urgent Maternal Protections for Nursing Mothers Act, aka the PUMP act.

76 *At the Equality Act hearing in 2019*: H.R. 5 Hearing Transcript, 156.

77 *Cambridge spokesperson Sophie White*: Timothy Bella, "Cambridge Dictionary

Updates Definition of 'Woman' to Include Trans Women," *Washington Post*, December 13, 2022, https://www.washingtonpost.com/dc-md-va/2022/12/13/woman-definition-cambridge-dictionary-transgender/.

78   *Dictionary.com made woman its Word of the Year:* "Dictionary.com's 2022 Word of the Year Is . . . ," Dictionary.com, December 13, 2022, https://www.dictionary.com/e/word-of-the-year/.

78   *the trans movement's fingerprints:* "Dictionary.com's 2022 Word of the Year."

80   *Justice Jackson thanked everyone:* Ketanji Brown Jackson, "Remarks by President Biden, Vice President Harris, and Judge Ketanji Brown Jackson on the Senate's Historic, Bipartisan Confirmation of Judge Jackson to be an Associate Justice of the Supreme Court," White House, April 8, 2022, accessed on September 14, 2023, https://www.whitehouse.gov/briefing-room/speeches-remarks/2022/04/08/remarks-by-president-biden-vice-president-harris-and-judge-ketanji-brown-jackson-on-the-senates-historic-bipartisan-confirmation-of-judge-jackson-to-be-an-associate-justice-of-the-supreme-court/.

80   *reaching the highest rank ever:* "The Order of Precedence of the United States of America," Department of State, revised on May 14, 2020, https://www.state.gov/wp-content/uploads/2020/05/2020-Order-of-Precedence-FINAL.pdf.

82   *"in the poetic words of Dr Maya Angelou":* Jackson, "Remarks."

4: Sex Is Good!

85   *His nakedness—in the nude sense:* John Paoletti, *Michelangelo's David: Florentine History and Civic Identity* (Cambridge: Cambridge University Press, 2015), 175–76.

86   *Renaissance painter and writer Giorgio Vasari:* "Michelangelo's David," Guide to the Accademia in Florence, accessed June 18, 2023, https://www.accademia.org/explore-museum/artworks/michelangelos-david/.

87   *Ginsburg's insistence that "'inherent differences'":* United States v. Virginia, 518 U.S. 515 (1996).

88   *graphic by statistician Kaj Tallungs:* Wikipedia, "Demographics of France," last modified June 19, 2023, https://en.wikipedia.org/wiki/Demographics_of_France.

88   *The pattern is essentially the same:* Covadonga Urquijo and Anne Milan, "Female Population," Statistics Canada, last modified November 30, 2015, https://www150.statcan.gc.ca/n1/pub/89-503-x/2010001/article/11475-eng.htm.

89   *Governments track birth rates and population numbers:* Executive Office of the President, Office of Management and Budget, "Statistical Programs of the United States Government," January 10, 2017, https://www.whitehouse.gov/wp-content/uploads/legacy_drupal_files/omb/assets/information_and_regulatory_affairs/statistical-programs-2017.pdf.

89   *On average, females also live five years longer:* E. Ginter and V. Simko "Women Live

Longer Than Men," *Bratislavske Lekarske Listy* [Bratislava Medical Journal] 114, no. 2 (2013): 45–49.

90 *"Population in More than 20 Countries to Halve by 2100"*: "Population in More than 20 Countries to Halve by 2100: Study," Al Jazeera, July 15, 2020, https://www.aljazeera.com/news/2020/7/15/population-in-more-than-20-countries -to-halve-by-2100-study.

91 *"China's Population Falls, Heralding a Demographic Crisis"*: Alexandra Stevenson and Zixu Wang, "China's Population Falls, Heralding a Demographic Crisis," *New York Times*, January 16, 2023, https://www.nytimes.com/2023/01/16 /business/china-birth-rate.html.

91 *Our fertility rate has dropped below 1.8*: Bill Chappell, "U.S. Birthrate Fell By 4% in 2020, Hitting Another Record Low," NPR, May 5, 2021, https://www.npr .org/2021/05/05/993817146/u-s-birth-rate-fell-by-4-in-2020-hitting-another -record-low.

91 *Lyman Stone of the Institute for Family Studies*: Lyman Stone, "5.8 Million Fewer Babies: America's Lost Decade in Fertility," Institute for Family Studies, February 3, 2021, https://ifstudies.org/blog/5-8-million-fewer-babies-americas-lost -decade-in-fertility.

92 *"a hospital . . . in Southwest China"*: Nicole Hong and Zixu Wang, "Desperate for Babies, China Races to Undo an Era of Birth Limits. Is It Too Late?" *New York Times*, February 26, 2023, https://www.nytimes.com/2023/02/26/world /asia/china-birth-rate.html.

92 *governments support measures linked indirectly to higher fertility rates*: Wikipedia, "Fertility Factor (Demography)," last modified February 16, 2023, https:// en.wikipedia.org/wiki/Fertility_factor_(demography).

93 *Dr. Hossam Fadel . . . explains*: Hossam E. Fadel, "The Islamic Viewpoint on New Assisted Reproductive Technologies," *Fordham Urban Law Journal* 30, no. 1 (2002): 147.

94 *kids affect men too*: Patty X. Kuo et al., "Fathers' Cortisol and Testosterone in the Days around Infants' Births Predict Later Parental Involvement," *Hormones and Behavior* 106 (2018): 28–34.

94 *At least some men*: Robin Edelstein et al., "Prospective and Dyadic Associations between Expectant Parents' Prenatal Hormone Changes and Postpartum Parenting Outcomes," *Developmental Psychobiology* 59, no. 1 (2016): 77–90; Emory Health Sciences, "How Dads Bond with Toddlers: Brain Scans Link Oxytocin to Paternal Nurturing," *Science Daily*, February 17, 2017, https://www.science daily.com/releases/2017/02/170217095925.htm.

95 *Human offspring are physically immature and dependent*: Kate Blackwood, "Prolonged Immaturity an Evolutionary Plus for Human Babies," *Cornell Chronicle*,

February 8, 2021, https://news.cornell.edu/stories/2021/02/prolonged-im maturity-evolutionary-plus-human-babies.

97   *Nonparents in the United States report:* Clay Routledge and Will Johnson, "The Real Story behind America's Population Bomb: Adults Want Their Independence," *USA Today,* October 12, 2022, https://www.usatoday.com /story/opinion/2022/10/12/why-americans-not-having-babies-low-birth -rate/8233324001/.

97   *In Portugal, for example:* "Which Is the Best Country to Be a Parent," Science Focus, BBC, January 23, 2022, https://www.sciencefocus.com/science/happi est-parents-country/.

98   *journalist Freddy McConnell: Seahorse: The Dad Who Gave Birth,* directed by Jeanie Finlay (London: Lobo Films/Andrea Cornwell, 2019).

98   *Sexual pleasure is:* Paul Abramson and Steven Pinkerton, *With Pleasure: Thoughts on the Nature of Human Sexuality* (New York: Oxford University Press, 1995), 8–10.

98   *The experience is broken down:* "Sexual Response Cycle," Cleveland Clinic, last reviewed March 8, 2021, https://my.clevelandclinic.org/health/articles/9119 -sexual-response-cycle.

99   *First comes desire:* S. B. Levine, "An Essay on the Nature of Sexual Desire," *Journal of Sex & Marital Therapy* 10, no. 2 (1984): 83–96.

99   *Then comes arousal:* "Sexual Response Cycle," Cleveland Clinic.

99   *Last is orgasm:* "Orgasm," Cleveland Clinic, last reviewed May 9, 2022, https:// my.clevelandclinic.org/health/articles/22969-orgasm.

99   *4–5 percent of us identify as bisexual:* Jeffrey M. Jones, "U.S. LGBT Identifica- tion Steady at 7.2%," Gallup, February 22, 2023, https://news.gallup.com /poll/470708/lgbt-identification-steady.aspx.

99   *British philosopher Kathleen Stock:* Kathleen Stock, "Lesbians Aren't Attracted to a Female 'Gender Identity.' We're Attracted to Women," Quillette, May 18, 2021, https://quillette.com/2021/05/18/lesbians-arent-attracted-to-a-female -gender-identity-were-attracted-to-women/.

100   *Veronica Ivy, formerly Rachel McKinnon:* Screenshot of tweet by Dr. Rachel Mc- Kinnon (@rachelvmckinnon), Twitter, September 30, 2019, 22:46.

100   *another version of this claim:* Caroline Lowbridge, "The Lesbians Who Feel Pres- sured to Have Sex and Relationships with Trans Women," BBC, October 26, 2021, https://www.bbc.com/news/uk-england-57853385.

101   *She's been bullied, deplatformed, and protested:* Gemma Parry, "Feminist univer- sity professor is told to get bodyguards and install CCTV at her home after balaclava-clad students demanded that she be sacked in vicious trans rights row," *Daily Mail,* October 10, 2021, https://www.dailymail.co.uk/news/article

-10077421/Feminist-university-professor-told-bodyguards-students-demanded
-sacked.html.

102 *political and cultural commentator Andrew Sullivan*: Andrew Sullivan, "Who Is
     Looking Out for Gay Kids?" Substack, April 8, 2022, https://andrewsullivan
     .substack.com/p/who-is-looking-out-for-gay-kids-a19.

104 *It can, among other things, reduce pain*: Michael Castleman, "8 Reasons
     Sex Improves Your Health," AARP, accessed June 6, 2023, https://www
     .aarp.org/relationships/love-sex/info-06-2011/sex-improves-men-health
     .html; "Orgasm," Cleveland Clinic.

104 *In Hinduism, sexual pleasure*: Rahul, "Four Purusharthas," Hinduism Facts, last
     edited June 17, 2021, https://www.hinduismfacts.org/four-purusharthas/.

104 *The millennia-old Kama Sutra*: Wendy Doniger and Sundhir Kakar, *Kamasutra*
     (Oxford: Oxford University Press, 2002), xi.

104 *Justice Harry Blackmun agreed*: Bowers v. Hardwick, 478 U.S. 186 (1986),
     J. Blackmun dissenting.

105 *the Supreme Court of India*: Johar v. Union of India, W.P. (Crl.) No. 76 of 2016,
     https://main.sci.gov.in/supremecourt/2016/14961/14961_2016_Judge
     ment_06-Sep-2018.pdf.

105 *the New World Encyclopedia*: "Beauty," New World Encyclopedia, accessed
     June 19, 2023, https://www.newworldencyclopedia.org/entry/Beauty.

107 *art historian Kenneth Clark*: Kenneth Clark, *The Nude: A Study in Ideal Form* (New
     York: Doubleday, 1956).

107 *the Venus de Milo*: Jane Ursula Harris, "The Role of the Copy," Believer, Au-
     gust 4, 2016, https://www.thebeliever.net/logger/2016-08-04-the-role-of-the
     -copy-2/.

107 *Contra Vasari on the David, Rodin proclaimed*: Auguste Rodin, "To the Venus De
     Milo," trans. Anna Seaton-Schmidt, *Art and Progress* 3, no. 2 (December 1911):
     409–13.

107 *Salvador Dalí made*: Branko van Oppen and Cindy Meijer, "Rediscovering the
     Venus de Milo," World History Encyclopedia, May 8, 2019, https://www
     .worldhistory.org/article/1377/disarming-aphrodite-rediscovering-the-venus
     -de-mil/.

108 *Taylor Hunt described Lansky's work*: Taylor Hunt, "Reimagining Venus de Milo
     in the Modern Age," ArtRKL, February 2023, https://www.artrkl.com/blogs
     /news/reimagining-venus-de-milo-in-the-modern-age.

108 *mosaics of female athletes in "bikinis"*: The mosaics are in the Room with Girls in
     Bikini in the Villa Romana del Casale, in Sicily. They're described, inter alia,
     in "An Overview of Roman Mosaics," Mused, accessed September 14, 2023,
     https://villaromana.mused.org/en/items/9309/room-with-girls-in-bikini.

110 *As Vice's Viola Zhou explained*: Viola Zhou, "Chinese Women Look at Eileen Gu

and Do Not See Themselves," Vice, February 15, 2022, https://www.vice.com/en/article/akv4v8/eileen-gu-china-women-privilege.

110 *soccer star Megan Rapinoe:* Kevin Baxter, "'The Best-Kept Secret in the State Department.' How Sports Help U.S. Diplomats," *Los Angeles Times,* January 29, 2023, https://www.latimes.com/sports/story/2023-01-29/sports-diplomacy-state-department.

111 *The David has been a repeat target:* Dan Kois, "An Interview with the School Board Chair Who Forced Out a Principal After Michelangelo's *David* Was Shown in Class," Slate, March 23, 2023, https://slate.com/human-interest/2023/03/florida-principal-fired-michelangelo-david-statue.html.

111 *The Art Newspaper reports:* Torey Akers, "Florence's Mayor Invites Florida Students and Their Former Principal to Experience the 'Purity' of Michelangelo's David," Art *Newspaper,* March 27, 2023, https://www.theartnewspaper.com/2023/03/27/florence-mayor-invites-floridians-michelangelo-david.

112 *two kinds of naked men in the piazza:* Paoletti, *Michelangelo's David,* 192, 196.

113 *"human potential in its most vital form":* Paoletti, *Michelangelo's David,* 198.

5: Sex Just Is (Like Age)

122 *Exploring the Biological Contributions to Human Health: Institute of Medicine, Does Sex Matter?*

122 *the National Academy of Medicine:* "About the National Academy of Medicine," National Academy of Medicine, 2023, https://nam.edu/about-the-nam/.

124 *this "stereotype" argument:* See, for example, Bostock v. Clayton County, 590 U.S. ___, (2020) and United States v. Virginia, 518 U.S. 515 (1996).

127 *Cognitively, there is almost complete overlap:* Linda Gottfredson, "Mainstream Science on Intelligence," *Wall Street Journal,* December 13, 1994.

129 *sex differences in heart structure and function:* Anna L. Beale et al., "Sex Differences in Cardiovascular Pathophysiology: Why Women Are Overrepresented in Heart Failure with Preserved Ejection Fraction," *Circulation,* July 9, 2018, https://www.ahajournals.org/doi/full/10.1161/CIRCULATIONAHA.118.034271.

130 *Miller explained in the* European Heart Journal: Virginia M. Miller, "Universality of Sex Differences in Cardiovascular Outcomes: Where Do We Go from Here?" *European Heart Journal* 41, no. 17 (May 1, 2020): 1697–99, https://doi.org/10.1093/eurheartj/ehaa310.

131 *Marking the twentieth anniversary:* Miller, "Universality of Sex Differences."

131 *reduce the granularity that matters:* Amy Eileen Hamm, "Teaching UBC Medical Scholars That Biological Sex Is a 'Colonial Imposition,'" Quillette, February 14, 2023, https://quillette.com/blog/2023/02/14/teaching-ubc-medical-students-that-biological-sex-is-a-colonial-imposition/.

131 *defining sex as biological sex:* "Sex Differences in Immunity in Health and

Disease," Gordon Research Conference, April 2, 2023, https://www.grc
.org/sex-differences-in-immunity-conference/2023/#:˜:text=One%20of%20
the%20most%20staggering,life%20course%20or%20through%20evolution
ary.

132   *"differences in overall immune function"*: Matthew Robert Schwartz and Marianne Berwick, "Sex Difference in Melanoma," *Current Epidemiology Reports* 6
      (June 15, 2019): 112–18, https://doi.org/10.1007/s40471-019-00192-7.

132   *differences are tied to age*: Sabra L. Klein and Katie L. Flanagan, "Sex Differences in Immune Responses," *Nature Reviews Immunology* 16, (2016): 626–38,
      https://doi.org/10.1038/nri.2016.90.

132   *females "mount more effective cellular"*: Carla Sebastián-Enesco and Gün R.
      Semin, "The Brightness Dimension as a Marker of Gender across Cultures
      and Age," *Psychological Research* 84 (2020): 2376–84, https://doi.org/10.1007
      /s00426-019-01213-2.

132   *The Covid pandemic brought the female-bias*: Azeen Ghorayshi, "Why Are Men More
      Likely to Die of Covid? It's Complicated," *New York Times*, January 19, 2022,
      https://www.nytimes.com/2022/01/19/health/covid-gender-deaths-men
      -women.html.

132   *greater immune response among women is "a blessing and a curse"*: Sarah Toy, "Covid-
      19 Vaccines and Rare Blood Clots: Are Women at Greater Risk?" *Wall Street
      Journal*, April 18, 2021, https://www.wsj.com/articles/covid-19-vaccines-and
      -rare-blood-clots-are-women-at-greater-risk-11618747200.

132   *The overall Covid pattern*: Klein and Flanagan, "Sex Differences in Immune Responses."

132   *80 percent of the patients are women*: Klein and Flanagan, "Sex Differences in Immune Responses."

133   *sex is also in our skin*: Sebastián-Enesco and Semin, "The Brightness Dimension."

133   *diagnosed with melanoma*: "Cate Campbell: MIA National Ambassador," Melanoma Institute Australia, September 21, 2021, https://melanoma.org.au
      /news/team/cate-campbell/.

133   *This includes Malawi*: Maurice Mulenga et al., "Epidemiological and Histopathological Profile of Malignant Melanoma in Malawi," *BMC Clinical Pathology* 19,
      no. 5 (April 2, 2019), https://doi.org/10.1186/s12907-019-0087-6.

134   *"biological sex is a fundamental factor in melanoma"*: Sebastián-Enesco and Semin,
      "The Brightness Dimension."

135   *she's added skin cancer prevention*: Melanoma Institute Australia.

6: *Sex Is Still a Problem (Like Race)*

138   *A caption accompanying a series of devastating photographs*: Al Shapiro et al.,
      "On the Front Lines of Poland's Makeshift Response to the Ukrainian

Refugee Crisis," NPR, March 15, 2022, https://www.npr.org/sections/pic
tureshow/2022/03/15/1086176563/on-the-frontlines-of-polands-makeshift
-response-to-the-ukrainian-refugee-crisis.

139 *listening to the BBC*: Quentin Sommerville, "Ukraine's Secret Weapon—the
Medics in the Line of Fire," BBC, November 15, 2022, https://www.bbc.com
/news/world-europe-63619240.

140 *Ukraine's surrogacy industry*: Isabel Coles, "Ukraine Is a World Leader in Surrogacy, but Babies Are Now Stranded in a War Zone," *Wall Street Journal*,
March 12, 2022, https://www.wsj.com/articles/ukraine-is-a-world-leader-in
-surrogacy-but-babies-are-now-stranded-in-war-zone-11647081997.

141 *the authors of "Sex Differences in the Human Brain"*: Alex DeCasien, Elisa Guma,
Siyuan Liu, and Armin Raznahan, "Sex Differences in the Human Brain,"
Biology of Sex Differences 13, no. 43 (2022), published online, doi:10.1186
/s13293-022-00448-w.

142 Domestic violence: "What Is Domestic Violence?" Office on Violence against
Women (OVW), U.S. Department of Justice, last updated March 17, 2023,
https://www.justice.gov/ovw/domestic-violence.

142 *victims of sex-based violence are female*: "Rape & Sexual Assault in the U.S.," National Organization for Women New York City, January 30, 2019, https://
nownyc.org/issues/get-the-facts-take-rape-seriously/.

143 *When they live on Indian reservations*: "Sexual Assault in Indian Country: Con-
fronting Sexual Violence," National Sexual Violence Resource Center, 2019,
https://www.nsvrc.org/sites/default/files/Publications_NSVRC_Booklets
_Sexual-Assault-in-Indian-Country_Confronting-Sexual-Violence.pdf.

143 *In the slums in Mumbai*: Caroline Criado Perez, *Invisible Women: Data Bias in a
World Designed for Men* (New York: Abrams Press, 2021), 50–51.

143 *UN secretary-general Ban Ki-moon*: "Ensuring Women's Access to Safe Toilets Is
'Moral' Imperative, Says Ban Marking World Day," United Nations Sustainable
Development Goals, November 19, 2014, https://www.un.org/sustainablede
velopment/blog/2014/11/ensuring-womens-access-safe-toilets-moral-impera
tive-says-ban-marking-world-day/.

143 *The perpetrators of sexual and domestic violence*: "Domestic Abuse is a Gendered
Crime," Women's Aid, accessed September 14, 2023, https://www.women
said.org.uk/information-support/what-is-domestic-abuse/domestic-abuse-is-a
-gendered-crime/.

144 *As I learned from Criado Perez*: Criado Perez, *Invisible Women*, 54–58.

144 *I no longer train at night*: Doriane Lambelet Coleman, "If Sex Matters Less," *Duke
Magazine*, August 8, 2019, https://alumni.duke.edu/magazine/articles/if-sex
-matters-less.

145 *"a major public health problem"*: "Violence against Women," World Health

Organization, March 9, 2021, https://www.who.int/news-room/fact-sheets /detail/violence-against-women.

146 *a sheriff in Mississippi*: Ilyssa Daly and Jerry Mitchell, "Sex Abuse, Beatings and an Untouchable Mississippi Sheriff," *New York Times*, April 11, 2023, https:// www.nytimes.com/2023/04/11/us/noxubee-mississippi-sheriffs-abuse.html.

146 *Iranian Revolutionary Guard*: Patrick Wintour, "Iranian Prosecutors Conceal Rape by Revolutionary Guards, Document Shows," *Guardian*, February 8, 2023, https://www.theguardian.com/world/2023/feb/08/iranian-prosecutors -concealed-by-revolutionary-guards-document-shows.

147 *75 percent of adult caregivers*: "Caregiver Statistics: Demographics," Family Caregiver Alliance, accessed June 6, 2023, https://www.caregiver.org/resource /caregiver-statistics-demographics/.

147 *According to DataUSA*: "Childcare Workers," DataUSA, accessed June 6, 2023, https://datausa.io/profile/soc/childcare-workers.

147 *Julia Haines of* U.S. News & World Report: Julia Haines, "Gender Reveals: Data Shows Disparities in Child Care Roles," *U.S. News & World Report*, May 11, 2023, https://www.usnews.com/news/health-news/articles/2023-05-11/gende-re veals-data-shows-disparities-in-child-care-roles.

148 *Caroline Criado Perez reports*: Criado Perez, *Invisible Women*, 40.

149 *A devastating article by Patrick Strickland*: Patrick Strickland, "Inside Owsley: America's Poorest White County," Al Jazeera, November 8, 2016, https://www.al jazeera.com/features/2016/11/8/inside-owsley-americas-poorest-white-county.

151 *double and even triple that of males*: A. W. Geiger and Leslie Davis, "A Growing Number of American Teenagers—Particularly Girls—Are Facing Depression," Pew Research Center, July 12, 2019, https://www.pewresearch.org/fact -tank/2019/07/12/a-growing-number-of-american-teenagers-particularly-girls -are-facing-depression/.

151 *57 percent of females in high school*: "CDC Report Shows Concerning Increases in Sadness and Exposure to Violence among Teen Girls and LGBQ+ Youth," Centers for Disease Control and Prevention, last reviewed March 9, 2023, https://www.cdc.gov/nchhstp/newsroom/fact-sheets/healthy-youth/sadness -and-violence-among-teen-girls-and-LGBQ-youth-factsheet.html#teen-girls.

152 *What's most alarming is that this number*: "CDC Report."

152 *"self-harm among 10-to-14-year-olds"*: "CDC Report."

152 *psychologist Jelena Kecmanovic*: Jelena Kecmanovic, "Why Tween Girls Especially Are Struggling So Much," *Washington Post*, August 8, 2022, https://www.wash ingtonpost.com/health/2022/08/08/tween-girls-mental-health/.

152 *boys are more likely to play computer games*: Elizabeth Englander and Meghan K. McCoy, "How to Help Teen Girls' Mental Health Struggles—6 Research-Based Strategies for Parents, Teachers and Friends," The Conversation, February 23,

2023, https://theconversation.com/how-to-help-teen-girls-mental-health-strug
gles-6-research-based-strategies-for-parents-teachers-and-friends-200052.

153 *the social media story*: Derek Thompson, "America's Teenage Girls Are Not
Okay," *Atlantic*, February 16, 2023, https://www.theatlantic.com/newsletters
/archive/2023/02/the-tragic-mystery-of-teenage-anxiety/673076/.

153 *the surgeon general of the United States*: "Social Media and Youth Mental
Health: The U.S. Surgeon General's Advisory," Office of the Surgeon General, May 23, 2023, https://www.hhs.gov/sites/default/files/sg-youth-mental
-health-social-media-advisory.pdf.

153 *In an interview with the* New York Times: Matt Richtel, Catherine Pearson, and
Michael Levenson, "Surgeon General Warns That Social Media May Harm Children and Adolescents," *New York Times*, May 23, 2023, https://www.nytimes
.com/2023/05/23/health/surgeon-general-social-media-mental-health.html.

154 *"thelarche by age eight"*: Jessica Winter, "Why More and More Girls Are Hitting
Puberty Early," *New Yorker*, October 27, 2022, https://www.newyorker.com
/science/annals-of-medicine/why-more-and-more-girls-are-hitting-puberty-early.

154 *Jessica Winter, reporting for the* New Yorker: Winter, "Why More and More Girls."

155 *As Winter put it*: Winter, "Why More and More Girls."

155 *Early puberty has both mid-term*: Tuck Seng Cheng, Ken K. Ong, and Frank M.
Biro, "Adverse Effects of Early Puberty Timing in Girls and Potential Solutions," *Journal of Pediatric & Adolescent Gynecology* 35, no. 5 (2022): 532–35,
https://doi.org/10.1016/j.jpag.2022.05.005.

156 *Reeves puts the matter bluntly*: Richard V. Reeves, "Redshirt the Boys," *Atlantic*,
September 14, 2022, https://www.theatlantic.com/magazine/archive/2022
/10/boys-delayed-entry-school-start-redshirting/671238/.

157 *Beyond secondary school, young men*: Richard V. Reeves, *Of Boys and Men: Why the
Modern Male Is Struggling, Why It Matters, and What to Do about It* (Washington,
DC: Brookings Institution Press, 2022), 16.

158 *"From a neuro-scientific perspective"*: Reeves, *Of Boys and Men*, 11.

158 *Writing in Quillette in 2021, the data scientist*: Vincent Harinam, "Mate Selection
for Modernity," Quillette, June 28, 2021, https://quillette.com/2021/06/28
/mate-selection-for-modernity/.

158 *Nicholas Eberstadt of American Enterprise Institute*: Nicholas Eberstadt, "The Idle
Army: America's Unworking Men," *Wall Street Journal*, September 1, 2016.

158 *Reeves adds that already before the pandemic*: Reeves, *Of Boys and Men*, 19.

158 *"one-two punch, of automation and free trade"*: Reeves, *Of Boys and Men*, 21–23.

161 *However we might rationalize our loneliness*: Daniel de Visé, "Most Young Men Are
Single. Most Young Women Are Not," *Hill*, February 22, 2023, https://the
hill.com/blogs/blog-briefing-room/3868557-most-young-men-are-single-most
-young-women-are-not/.

161 *Like the costs of men's educational and workplace losses*: Ian Taylor, "How Loneliness Is Killing Men," Science Focus, BBC, November 11, 2022, https://www .sciencefocus.com/science/how-loneliness-is-killing-men/.

161 *One study found that "long-term social isolation"*: Andrea M. Tilstra, Daniel H. Simon, and Ryan K. Masters, "Trends in 'Deaths of Despair' Among Working-Aged White and Black Americans, 1990–2017," *American Journal of Epidemiology* 190, no. 9 (2021): 1751–59, https://doi.org/10.1093/aje/kwab088.

162 *men are more likely to suffer deaths of despair*: Jed Diamond, PhD, "Deaths of Despair: Are Males More Vulnerable?" Medium, January 21, 2022, https://medium.com/equality-includes-you/deaths-of-despair-are-males-more-vulnerable -f43047d9125d.

162 *"far higher" among American Indians*: Rhitu Chatterjee, "Native Americans Left Out of 'Deaths of Despair' Research," NPR, February 1, 2023, https://www .npr.org/sections/health-shots/2023/02/01/1152222968/native-americans -left-out-of-deaths-of-despair-research.

162 *Everett Rhoades, the first Native American director*: Everett R. Rhoades, "The Health Status of American Indian and Alaska Native Males," *American Journal of Public Health* 93, no. 5 (2003): 774–78, https://ajph.aphapublications.org /doi/full/10.2105/AJPH.93.5.774.

163 *Vladimir Putin adopted this new law*: Ivan Nechepurenko, Neil MacFarquhar, and Vjosa Isai, "Russia Moves to Make Draft Evasion More Difficult," *New York Times*, April 11, 2023, https://www.nytimes.com/2023/04/11/world/europe/russia -military-draft-ukraine.html?te=1&nl=the-morning&emc=edit_nn_20230412.

7: The (Un)Lawfulness of Regulating on the Basis of Sex and Gender

167 *"famously tiny"*: Vanessa Friedman, "Ruth Bader Ginsburg's Lace Collar Wasn't an Accessory, It Was a Gauntlet," *New York Times*, September 23, 2020, https:// www.nytimes.com/2020/09/20/style/rbg-style.html.

167 *But as Sharron Frontiero*: Sharron Cohen and Nathan Cohen, "'It's Okay to Be a Hero': Remembering Justice Ginsburg's Words," StoryCorps, originally aired December 18, 2020, on NPR's *Morning Edition*, https://storycorps.org/stories /its-okay-to-be-a-hero-remembering-justice-ginsburgs-words/.

170 *New York State Woman Suffrage Association*: "Arguments for and against Suffrage," Women & the American Story, New-York Historical Society, https:// wams.nyhistory.org/modernizing-america/woman-suffrage/arguments-for -and-against-suffrage/#.

170 *Crystal Eastman, a cofounder of the ACLU*: Crystal Eastman, "'Now We Can Begin': What's Next? Beyond Woman Suffrage," Women's History Guide, accessed August 3, 2023, https://womenshistory.info/now-can-begin-whats-next -beyond-woman-suffrage/.

170 *women are like men*: These examples come from the three sex classification cases the Supreme Court decided in the early to mid-twentieth century. Applying the rational basis test, the Court upheld state labor laws prohibiting employers from making women work for more than ten hours a day, Muller v. Oregon, 208 U.S. 412 (1908); prohibiting women from working as bartenders unless their husband or father owned the drinking establishment, Goesaert v. Cleary, 335 U.S. 464 (1948); and requiring women to opt in to the jury pool if they wanted to serve as jurors, Hoyt v. Florida, 368 U.S. 57 (1961).

172 *waitresses but not bartenders*: Justice Frankfurter wrote the majority opinion in Goesaert v. Cleary, 335 U.S. 464, 465-66 (1948) (finding reasonable a state law denying females bartender licenses).

172 *In Korematsu, the Court held*: 323 U.S. 214, 216 (1944).

174 *winning strategy in Brown*: Tejai Beulah Howard, "How Pauli Murray Master-minded Brown v. Board," *Black Perspectives* (blog), October 13, 2022, https://www.aaihs.org/how-pauli-murray-masterminded-brown-v-board/.

174 *Commission on the Status of Women*: United States President's Commission on the Status of Women Records (#400), Collection Overview, John F. Kennedy Presidential Library and Museum, https://www.jfklibrary.org/asset-viewer/archives/USPCSW#admininfo.

174 *Murray wrote a memorandum*: Pauli Murray, "A Proposal to Reexamine the Applicability of the Fourteenth Amendment to State Laws and Practices Which Discriminate on the Basis of Sex Per Se," December 1, 1962, Box 8, Folder 62, United States President's Commission on the Status of Women Records, 1961–1963, Schlesinger Library, Radcliffe Institute for Advanced Study, Harvard University.

174 *call the "remedies" in the "arsenal"*: Brief for Appellant, Reed v. Reed, 404 U.S. 71 (1971) (No. 70-4), 12, https://socialchangenyu.com/wp-content/uploads/2019/08/1970-Reed-v-Reed-Brief-for-Appellant.pdf.

174 *"we should interpret the text literally"*: Olivia B. Wasman, "In Previously Unseen Interview, Ruth Bader Ginsburg Shares How Legal Pioneer Pauli Murray Shaped Her Work on Sex Discrimination," *Time*, October 20, 2020, https://time.com/5896410/ruth-bader-ginsburg-pauli-murray/.

175 *"The central constitutional issue"*: Murray Memorandum, 6.

175 *the sociological case against sex discrimination*: Murray Memorandum, 11 and notes 18–23.

175 *Murray's memo made it into Ginsburg's hands*: Philippa Strum, "Pauli Murray's Indelible Mark on the Fight for Equal Rights," ACLU, June 24, 2020, https://www.aclu.org/issues/womens-rights/pauli-murrays-indelible-mark-fight-equal-rights.

176 *"we were standing on her shoulders"*: Wasman, "In Previously Unseen Interview."

176 *"full human personalities"*: Brief for the Appellant, Reed, 404 U.S. 71 (1971) (No. 70-4).

177   *vying for the assignment*: Brief for Appellant, Reed, 404 U.S. 71 (1971) (No. 70-4).

182   Craig *effectively required policymakers*: Craig v. Boren, 429 U.S. 190 (1976).

183   *He also dismissed as "myth"*: Dothard v. Rawlinson, 433 U.S. 321 (1977).

184   Personnel Administrator of Massachusetts v. Feeney: Personnel Administrator of Massachusetts v. Feeney, 442 U.S. 256 (1979).

184   *Because* Feeney *didn't involve a sex classification*: Mississippi University for Women v. Hogan, 458 U.S. 718, 724–26 (1982) (internal citations omitted).

186   *"excluding men from the School of Nursing"*: Hogan, 458 U.S. at 731.

187   City of Richmond v. J. A. Croson Company: City of Richmond v. J. A. Croson Co., 488 U.S. 469 (1989) (internal citations omitted).

189   J.E.B. v. Alabama: J.E.B. v. Alabama ex rel. TB., 511 U.S. 127 (1994).

189   *Justice Harry Blackmun insisted*: J.E.B., 511 U.S. at 143, n. 11.

190   *Scalia wrote a dissenting opinion*: J.E.B., 511 U.S. at 156–58.

192   United States v. Virginia: 518 U.S. 515 (1996).

192   *the Clinton administration filed a brief*: Brief for United States in No. 94-2107, 16. Scalia notes that this position "was in flat contradiction of the Government's position below, which was, in its own words, to 'stat[e] *unequivocally* that the appropriate standard in this case is "intermediate scrutiny."'" 2 Record, Doc. No. 88, 3 (emphasis added). 518 U.S. at 571.

192   *Ginsburg's "constitutional vision"*: Neil S. Siegel, "'Equal Citizenship Stature': Justice Ginsburg's Constitutional Vision," *New England Law Review* 43, no. 799 (2009).

192   *As Ginsburg herself put it*: United States v. Virginia, 518 U.S. at 532 (emphasis added).

192   *Ginsburg's VMI opinion*: 518 U.S. at 523–34 (all internal citations omitted).

194   *"no 'exceedingly persuasive justification'"*: 518 U.S. at 555–56.

194   *prompted a dissenting Scalia*: 518 U.S. at 567–603 (Scalia, J., dissenting).

194   *expert testimony about "gender-based developmental differences"*: 518 U.S. at 549, 585 (Scalia, J., dissenting).

195   *"the Court's unease" with "the consequences"*: 518 U.S. at 595 (Scalia, J., dissenting).

196   *She celebrated her first Eucharist*: Episcopal Diocese of North Carolina, "Pauli Murray, Biographical Timeline," accessed June 13, 2023, https://www.epis dionc.org/pauli-murray/.

8: The Politics of Sex and Gender

198   *as Justice Ginsburg would have allowed*: United States v. Virginia, 518 U.S. 515, 533-34 (1996).

200   *"experts in women"*: Toward the end of chapter 3 I describe a statement by the emissary from the National Women's Law Center to the 2019 Equality Act hearing in the House Judiciary Committee in which she makes this claim. For the surrounding discussion see that chapter.

200 *Capitalizing on this familial fracture*: Madeleine Kearns, "The Equality Act Is a Time Bomb," *National Review*, May 20, 2019, https://www.nationalreview .com/corner/the-equality-act-is-a-time-bomb/.

200 *"Pass the Equality Act, but Don't Abandon Title IX"*: Doriane Lambelet Coleman, Martina Navratilova, and Sanya Richards-Ross, "Pass the Equality Act, but Don't Abandon Title IX," *Washington Post*, April 29, 2019, https://www.wash ingtonpost.com/opinions/pass-the-equality-act-but-dont-abandon-title-ix/201 9/04/29/2dae7e58-65ed-11e9-a1b6-b29b90efa879_story.html.

201 *"A Victory for Female Athletes Everywhere"*: Doriane Lambelet Coleman, "A Victory for Female Athletes Everywhere," Quillette, May 3, 2019, https://quillette .com/2019/05/03/a-victory-for-female-athletes-everywhere/.

203 *Chase Strangio put it to Martina and me*: Chase Strangio, "How Many Trans Kids Will Die While We Await This Mythic Trans Sports Takeover?" Medium, January 22, 2021, https://chase-strangio.medium.com/how-many-trans-kids-will -die-while-we-await-this-mythic-trans-sports-takeover-8a4ab53324da.

205 *participation, not the podium*: Michael J. Joyner, Donna Lopiano, and I discuss this claim in detail on pages 117–19 of our law review article "Reaffirming the Sports Exception to Title IX's General Nondiscrimination Rule," *Duke Journal of Gender Law & Policy* 27 (2020): 69–134.

208 *The word transgender may be new*: For a recent historical work on the subject see, for example, Kit Heyam's book *Before We Were Trans: A New History of Gender* (New York: Seal Press, 2022).

208 *The Smithsonian Institution*: Katherine Ott, "The History of Getting the Gay Out," National Museum of American History, November 15, 2018, https:// americanhistory.si.edu/blog/getting-gay-out.

210 *Michelle Goldberg of the* New York Times: *The Ezra Klein Show*, "Michelle Goldberg Grapples with Feminism After Roe."

211 *President Barack Obama's celebratory remarks*: Ben Jacobs, "'Love Is Love': Obama Lauds Gay Marriage Activists in Hailing 'a Victory for America,'" *Guardian*, June 26, 2015, https://www.theguardian.com/us-news/2015/jun/26/obama -gay-marriage-speech-victory-for-america.

211 *the ACLU's Q&As on California law*: "Rights of Transgender and Nonbinary People in Gyms and Spas," ACLU Southern California, accessed September 14, 2023, https://www.aclusocal.org/en/know-your-rights/rights-transgender-and -nonbinary-people-gyms-and-spas-0.

212 *"Always brand sanitary pads"*: Heather Murphy, "Always Removes Female Symbol from Sanitary Pads," *New York Times*, October 22, 2019, https://www.nytimes .com/2019/10/22/business/always-pads-female-symbol.html.

212 *add pronouns to their signature lines*: Staci Zaretsky, "Biglaw [sic] Firm Encourages All Employees to Use Gender Pronouns in Email Signatures," Above the Law,

November 22, 2021, https://abovethelaw.com/2021/11/biglaw-firm-encour
ages-all-employees-to-use-gender-pronouns-in-email-signatures/.

212   *"She Killed Two Women"*: Rebecca Davis O'Brien and Ali Watkins, "She Killed Two Women. At 83, She Is Charged with Dismembering a Third," *New York Times,* March 10, 2022.

214   *Michelle Goldberg of the* New York Times: *The Ezra Klein Show,* "Michelle Goldberg Grapples with Feminism."

215   *a 2023 essay titled "Gay Rights and the Limits of Liberalism"*: Andrew Sullivan, "Gay Rights and the Limits of Liberalism—How the Far Left Broke the Gay Settlement. And Reignited Homophobia," Weekly Dish, June 23, 2023, https://andrewsullivan.substack.com/p/gay-rights-and-the-limits-of-liberalism-be4.

217   *Florida's "Parental Rights in Education" law:* Eesha Pendharkar, "Florida Just Expanded the 'Don't Say Gay' Law. Here's What You Need to Know," *Education Week,* April 19, 2023, https://www.edweek.org/policy-politics/florida-just-expanded-the-dont-say-gay-law-heres-what-you-need-to-know/2023/04.

217   *Florida governor Ron DeSantis:* Jaclyn Diaz, "Florida's Governor Signs Controversial Law Opponents Dubbed 'Don't Say Gay,'" NPR, March 28, 2022, https://www.npr.org/2022/03/28/1089221657/dont-say-gay-florida-desantis.

222   *Trans Journalists Association* Style Guide: *Stylebook and Coverage Guide,* Trans Journalists Association, accessed July 1, 2023, https://www.transjournalists.org/style-guide/.

223   *Fae Johnstone, an influential Canadian trans activist:* screenshot of tweet by Fae Johnstone (@FaeJohnstone), Twitter, July 12, 2021.

223   *"A strategy of the far right re: trans people"*: screenshot of tweet by Fae Johnstone (@FaeJohnstone), Twitter, March 15, 2023, 08:26 a.m.

223   *ultraconservatives like Matt Walsh:* Here's the full text of this particular tirade, which appeared on Walsh's Twitter blog with the caption "A Heartfelt Message to Dylan Mulvaney":

Dylan, if that is the most attractive that you will ever look then I don't even want to imagine what you'll look like when you're at your ugliest. You do not pass as an attractive woman or as a woman at all. Even with fifty pounds of makeup and plastic surgery and clever lighting tricks. Even then you still cannot escape what you really are and what you will always be. You have successfully shed whatever parts of you were masculine, perhaps. At least on the surface, no one would ever describe you as masculine or manly. So you've got that going. But your femininity-quotient has not increased at a rate commensurate to the loss of your masculinity. You may not be masculine but you also aren't feminine. Instead, you are weird and artificial; you are manufactured and lifeless; you are unearthly and eerie. You are like some kind of human deepfake. That's what you are. You are a man deprived of all the best qualities of men but without any of the best qualities of women. Even your personality is contrived. Everything about you is fake. Nothing about you rings true. Nobody buys the act. You will never be accepted as a woman by anyone. Even the people that pretend to accept you as a woman are only pretending because they're afraid of being lectured if they don't. Or because they want to use you as a platform to virtue-signal. But everyone who looks at you will see something pitiable and bizarre; something utterly unfeminine in every way. You will never be able to actually have the identity that you're trying to appropriate. Nor will you ever be able to fully escape the identity that you're fleeing. The best that you can hope for is some kind of limbo; the worst of all worlds. And yet, even in that limbo state, you will still be a man. Just not one that any of us can respect or take seriously. But other than that, champ, you're doing great.

Screenshot of tweet by Matt Walsh (@MattWalshBlog), Twitter, February 14, 2023, 17:06, https://twitter.com/MattWalshBlog/status/1625617441219813 409?lang=en.

226 *the* Chronicle of Higher Education: Tom Bartlett, "The Essay That Prompted an Editorial Revolt," *Chronicle of Higher Education*, March 8, 2022, https://www.chronicle.com/article/the-essay-that-prompted-an-editorial-revolt.

227 *Order of the British Empire*: "University of Sussex Professor Awarded OBE," press release, University of Sussex, January 5, 2021, https://www.sussex.ac.uk/broadcast/read/54254.

228 *"The Sami people"*: David Robson, "There Really Are 50 Eskimo Words for 'Snow,'" *Washington Post*, January 14, 2013, https://www.washington post.com/national/health-science/there-really-are-50-eskimo-words-for-snow/2013/01/14/e0e3f4e0-59a0-11e2-beee-6e38f5215402_story.html.

229 *"marriage by capture"*: Doriane Lambelet Coleman, "Individualizing Justice Through Multiculturalism: The Liberals' Dilemma," *Columbia Law Review* 96 (June 1996): 103.

229 *Using the word* gender *to mean* sex: Criado Perez, *Invisible Women*, 7.

229 *an uproar at Johns Hopkins*: Matt Lavietes, "Johns Hopkins Pulls 'Lesbian' Definition after Uproar over Use of 'Non-Men' instead of 'Women,'" NBC News, June 14, 2023, https://www.nbcnews.com/nbc-out/out-news/johns-hopkins-pulls-lesbian-definition-uproar-use-non-men-instead-wome-rcna89307.

9: A Commonsense Approach

234 *namesake for a multi-day festival in 1969*: "Woodstock Music Festival Site," National Park Service, accessed August 31, 2023, https://www.nps.gov/places/woodstock-music-festival-site.htm.

238 *"Sex is a true binary"*: Richard Dawkins, "Why Biological Sex Matters," *New*

Statesman, July 26, 2023, https://www.newstatesman.com/ideas/2023/07 /biological-sex-binary-debate-richard-dawkins.

238   *"There are two sexes in mammals, and that's that"*: Dawkins, "Why Biological Sex Matters."

252   *Katie Ledecky*: Dave Sheinin, "How Katie Ledecky Became Better at Swimming than Anyone Is at Anything," *Washington Post*, June 24, 2016, https:// www.washingtonpost.com/sports/olympics/how-katie-ledecky-became-better -at-swimming-than-anyone-is-at-anything/2016/06/23/01933534-2f31-11e6 -9b37-42985f6a265c_story.html.

259   *transmedicalism is anti-trans hate*: Mey Rude, "Hunter Schafer Addressed That Anti-Nonbinary Instagram Like," *Out*, September 6, 2022, accessed September 13, 2023 https://www.out.com/celebs/2022/9/06/hunter-schafer -addresses-anti-nonbinary-instagram.

259   *Joanna Harper is helpful here*: Harper discusses her experience in the larger scientific and sociological contexts in her book *Sporting Gender: The History, Science, and Stories of Transgender and Intersex Athletes* (Lanham, MD: Rowman & Littlefield, 2019).

260   *Harvey Marcelin, aka Marceline Harvey*: Rebecca Davis O'Brien and Ali Watkins, "How Did a Two-Time Killer Get out to Be Charged Again at Age 83?" *New York Times*, August 2, 2022, https://www.nytimes.com/2022/07/30/nyre gion/how-did-a-two-time-killer-get-out-to-be-charged-again-at-age-83.html.

261   Vindication *is a leg in the centuries-long relay*: Mary Wollstonecraft, *A Vindication of the Rights of Woman* (London: Penguin Classics, 1988), 79.

262   *"it is a noble prerogative!"*: Wollstonecraft, *Vindication*, 80.

262   *"Women, I allow"*: Wollstonecraft, *Vindication*, 139.

263   A Sculpture for Mary Wollstonecraft: Mark Brown, "Mary Wollstonecraft Finally Honoured with Statue after 200 Years: 'Mother of Feminism' Commemorated by Maggi Hambling Sculpture in North London," *Guardian*, November 9, 2020, https://www.theguardian.com/books/2020/nov/10/mary-wollstonecra ft-finally-honoured-with-statue-after-200-years.

263   *"I don't get why she's naked"*: Mark Brown, "Mary Wollstonecraft Statue Becomes One of 2020's Most Polarizing Artworks," *Guardian*, December 25, 2020, https://www.theguardian.com/artanddesign/2020/dec/25/london-mary -wollstonecraft-statue-one-of-2020s-most-polarising-artworks.

264   *I encourage you to read* Vindication *for yourself*: Wollstonecraft, *Vindication*, 81–82, 105, 129–30.

Index

Abdullah, Crown Prince of Saudi Arabia, 50
abortion, abortion rights, xxi, 43, 60, 64, 92,
    140, 163–64, 168, 217
  conservative women's argument against,
    236–37
  Dobbs decision and, 37, 136–38, 168
  transgender rights in debate on, 65, 207
  "trigger" laws on, 138
Above the Law (blog), 52–53
academic freedom, 220, 221, 226
Académie Française, 7–8
Adams, Drew, 18, 71
Adams v. School Board of St. Johns County, Florida,
    71–72
Adarand Constructors v. Peña, 189, 191–92, 193
Adichie, Chimamanda Ngozi, 261
affirmative action, 187, 191–92, 221
Afghanistan, 141
African Americans, 55–56, 168, 211, 233, 257,
    258
  civil rights movement and, 172, 173, 199
  discrimination against, *see* racism, racial
    discrimination
  see also black women; slavery
aggression, in males, 20, 43, 49, 127, 144
aging, 42–43, 91, 95
  blood pressure and, 130
  in brain, 126
  hormone levels and, 21–22
  immune system and, 132
  melanoma and, 134–35
Air Force, U.S., 179
Alaska Native people, 162
Alcott, Louisa May, 74
Alexandros of Antioch, 107
Algorithmic Beauty (Lansky), 108

Al Jazeera, 90, 149
Alliance Defending Freedom, Center for Life
    at, 236
Alzheimer's disease, 128, 129
amateur sports, 115, 119, 121
American Civil Liberties Union (ACLU), 61,
    64–65, 170, 176, 196, 203, 204, 211, 219,
    221, 225
American Enterprise Institute, 158
American Indians, 143, 162
Ancestry.com, 258
Angelou, Maya, 82
Anheuser-Busch, 214
anti-doping programs, xiii–xiv
anxiety, 21, 48, 128, 145, 151, 154, 155, 218, 249
Arcidiacono, Peter, 221
Arizona, 187
Arnold, Art, 128
arousal, in sexual response cycle, 99
art, 85–86, 107–8, 111, 112–13, 241, 263
  athletic body in, 108–10, 263
Art Newspaper, 111
Associated Press, 240
Association for Intercollegiate Athletics for
    Women, xiii
Astell, Mary, 49
Athlete Ally, 202
athletic body:
  beauty of, 108–10, 118
  see also female athletes
Atlantic, 151
Auchus, Richard J., 23
Australia, 133

Ban Ki-moon, 143
Bannon, Steve, 203

Barlow, Irene, 196
Barrett, Amy Coney, 206-7
bathrooms:
  debate over transgender use of, 18, 66, 71-72, 242-43, 244-46
  gender-neutral, xiii, 66, 67, 242, 244
BBC, 139
beauty, 234
  of athletic body, 108-10, 118, 263
  definition of, 105-6
  efforts in taking down hierarchies of, 111-12
  in gender-bending expressions, 241-42
  in male and female bodies, 85, 86, 106-13, 140, 232, 241, 263-64
Bennett, Katie, 149
Biden, Joe, xix, 73, 75, 211, 213-14, 219
bimodal development, 15, 20
biological sex, xiv, 4, 14-15, 21, 25, 63, 100, 113, 121, 141, 188, 198, 231
  as binary, 7, 8-10, 11, 22, 24-25, 66-67, 69-70, 101-2, 103, 120, 199, 206, 214, 215, 238-39, 255-56
  definition of, xii, 7-10, 40, 75-76, 123, 131-32, 239
  differences due to, *see* sex differences
  differential diagnosis for, 13, 239
  erasure of, progressive advocacy of, xii, xv, xviii, xix, 29, 31, 44, 63, 64-65, 67-68, 70-71, 72, 73, 75-76, 87, 198, 199, 203, 207, 211, 213, 214-16, 222, 227, 240, 259
  gender identity vs., 25-26, 29, 33, 40-41, 45, 47, 65, 123, 131, 239-40, 242, 252, 253
  identifying and distinguishing of, 26-27
  law's recognition of, 40, 44-45, 47 *see also* sex classifications
  procreation and, 88-89, 90, 91-92, 93-95, 214
  recognition of, in traditional faction of feminism, 64-65, 66-67, 68, 87, 101, 176-77, 214-15, 222, 223-25, 226, 227, 261
  sex characteristics and, *see* sex characteristics
  sex discrimination as based on, xviii, 64
  sexual pleasure and, 98-105, 113
  "social construction" view of, 68-72, 199
  sports eligibility standards based on, *see* sex-based eligibility standards
  see also sex classifications
Biology of Sex Differences, 125
birth certificates, 46-47, 67, 207
birth control, 70-71, 92, 137, 237

Birth of the Cool, The (album), 107
birth rates, 89, 90
bisexuality, 20, 65, 99, 103, 241
Bishop, Barney, III, 111
Blackburn, Marsha, 59, 60, 62, 63, 77, 78, 79, 261
Blackmun, Harry, 104, 180, 189, 190
Black's Law Dictionary, 43-44, 45
black women, 174, 196, 197
  dropping age of puberty in, 154
  as historically sidelined in women's rights movement, 60, 173
  Jackson's Supreme Court confirmation and, 56, 58, 59, 80-81
  racist stereotypes of, 57, 59-60, 62
  in sex-based eligibility standards debate, 60-62
  in slavery, 51, 56, 57, 59
  social masculinizing of, 28, 59
  as treated in U.S. law, 56, 57
  treatment of white women vs., 59-60
blood pressure, 130
Bluette (author's mother), 133-34, 135, 232-34, 235
body dysphoria, 94-95
body image, 116, 153
body shaming, 112, 264
book bans, 43, 217, 219
Booker, Cory, 58
Booneville, Ky., 149
Boroditsky, Lera, 72-73
Bostock v. Clayton County, 44, 52, 209
Botswana, 143, 145
Botton, Wesley, 6
Bowers v. Hardwick, 104
Bowie, Ariana, 137
Bowie, Tori, 137
Boylan, Jennifer Finney, 71
Bradley, Joseph, 37-40, 42, 43, 52, 53, 54, 192, 206, 222
Bradwell, Myra, 34, 35-37, 40, 47, 52, 53-54, 137, 168, 171, 176
Bradwell v. State of Illinois, 36-40, 42, 52, 53-54, 192
brain, brain development, 128
  in adolescence, 21, 48, 128, 152-53, 154-55, 157-58, 248
  antisexist approach to research on, 125
  gender identity and, 25, 70, 71-72, 247
  plasticity of, 19, 128
  sex differences in, 19-20, 21, 48, 49-50, 125, 126-29, 130, 152, 154-55, 157, 248

brain tumors, 129
Brandeis University, 196
breast cancer, 122, 155, 196
breastfeeding, 17, 51, 76, 93, 94
Breitbart News, 203–4
Brennan, William, 180, 181, 183
Brisbane, Australia, 116
Brookings Institution, 156
Brown, Mark, 263
Brown, Mike, 129
Brown v. Board of Education, xvi, 172, 173, 174, 236
Bryant, Kobe, 241
Budapest, Hungary, 114, 116, 119, 121
Bud Light, 213–14
Burger, Warren, 178, 180, 189
Bush, George W., 50
Buttigieg, Pete and Chasten, 236

Calamia, Cal, 253
California, 41, 145, 209, 211, 233
California, University of, Los Angeles (UCLA), 225
California Health and Safety Code, 46
Cambridge Dictionaries, 258–59
Cambridge English Dictionary, 77–78, 79
Campbell, Cate, 109, 114–15, 116, 117–21, 260
  melanoma diagnosis of, 133, 135
Campbell, Naomi, 241
cancer, 122, 132, 133–35, 155, 161, 196
caregiving norms, 146–51
Carpenter, Matthew, 36–37, 168, 176
"Castelike Status of Women and Negroes," 175
"Caster Semenya Is Being Forced to Alter Her Body to Make Slower Runners Feel Secure in Their Womanhood" (Strangio), 61
Centers for Disease Control and Prevention (CDC), 151
Centre de Procréation Medicalement Assistée, 93
Chalamet, Timothée, 241–42
Chastain, Brandi, xvii
ChatGPT, xxi, 79–80, 256
Chicago Legal News, 35–36, 54
China, 91, 92, 110
chromosomes, 8, 9, 10–14, 21, 23, 25, 40, 69, 70, 71, 72, 121, 130, 131, 224, 231, 238, 239
  as causing sex differences in brain, 19, 20
  disease pattern differences and, 128, 135

sex-based eligibility standards and, 30, 31, 115, 253
Chronicle of Higher Education, 226
City of Richmond v. J. A. Croson Company, 187
Civil discourse, 11, 78,263
Civil Rights Act (1964), 44, 74–75, 209
civil rights movement, 172, 173, 199, 215
Civil War, U.S., 45, 57, 167, 168
Clark, Kenneth, 107
Clarke, Edward, 49
Clinton administration, xiv, 192
Cohen, David, 178–79
Cole, Chandler, 139–40
Coles, Isabel, 140
Collins, Eliza, 169
Collins, Nina Lorez, 95
Collins, Robert, 169
Combat Exclusion (podcast), 139
Commission on the Status of Women, 174
complete androgen insensitivity syndrome (CAIS), 23, 24, 26, 257
congenital adrenal hyperplasia (CAH), 23–24, 26
Congress, U.S., xvii, xix, 171
  see also House of Representatives, U.S.; Senate, U.S.
conscription, military, 44–45, 136, 139, 141, 162, 163, 232
Constitution, U.S., 36, 37, 55, 167, 172, 210
  see also specific amendments
contraception, 71, 92, 137, 237
contract law, 38, 169
conversion therapy, 209
Cornell University, xiii, 171
Court of Arbitration for Sport (CAS), xiv, 30–33, 61
Covid pandemic, 52, 132, 146, 151, 152, 225
Craft, Ellen, 56–57, 169, 233
Craft, William, 233
Craig v. Boren, 181–83, 188
Criado Perez, Caroline, 141, 144, 148, 229, 264
Crisman, Johanna, 139–40
cross-sex hormones, 17, 18, 26, 66, 67, 71, 73, 231, 246–47, 248, 249–50, 253, 259–60
Czechoslovakia, 4, 202

Dalí, Salvador, 107
Dana Farber Cancer Institute, 134
Darity, William A. "Sandy," 221
Darwin, Charles, 9
DataUSA, 147

David (Michelangelo), 85–86, 107, 108, 111, 112–13, 209, 241, 263
Davis, Miles, 107
Dawkins, Richard, 238
DeCasien, Alex, 125, 127
de la Cruz, Ben, 138
Democratic Party, Democrats, xi, xiv, xv, 198, 200, 203, 204, 209, 216
depression, 21, 48, 128, 145, 151, 153, 154, 155, 161, 218, 249
de Reuse, Willem, 228–29
DeSantis, Ron, xx, 217–18
desire, in sexual response cycle, 99
de Varona, Donna, 205, 225
de Visé, Daniel, 161
Dialog, 96
Dictionary.com, 78–79
dimorphism, 8–10, 14, 15, 17, 18, 22, 24, 25, 26
Dine, Jim, 107
Discobolus (sculpture), 108
disease, sex differences and, 43, 123, 128, 129, 131, 132, 134–35, 161
disorders of sex development (DSD), 23–26, 29–32, 65, 70, 121, 238, 239, 253, 256, 257
Dobbs v. Jackson Women's Health Organization, 37, 136–38, 168, 216–17
Does Sex Matter? (IOM report), 122, 123, 131
domestic violence, 142, 143, 144, 145, 146, 162, 226
"Don't Say Gay" law, 217–18, 219
Dothard v. Rawlinson, 183
Douglas, William O., 180
drafts, military, 44–45, 136, 139, 141, 163, 232
Due Process Clause, 168, 209
Duke University, 11, 129, 220–21, 226

early onset gender dysphoria, 25, 247
Eastman, Crystal, 170, 171
eating disorders, 145, 153, 155
Eberstadt, Nicholas, 158, 159
education, 43, 57, 141, 156, 159–60, 262
assault on academic freedom in, 220–21, 226
banning of LGBTQ issues in, xx, 217–18, 219
males as underperforming in, 156–58, 161
racial segregation in, 173
single-sex, 194
eggs, 10, 11, 12, 13–14, 17
Eisenstadt v. Baird, 137

elite sports, xx, 114, 202, 205, 251
amateur, youth and community sports vs., 115, 119
sex-based eligibility standards for, 28, 114–15, 119–21, 124, 238–39, 250, 253–54
sex equality in, 115–16, 119–20
Emancipation Proclamation, xvi, 230
Enfants de la Ville, Les (documentary), 233
Episcopalian Church, 196–97
Equality Act (H.R. 5), xi, xii–xiii, xiv–xx, 74–77, 198–201, 210, 212, 230
female sports and, xiii, xvi–xvii, 200–202, 204
presented as "good for women," xix, 76–77, 199–200
redefining "sex" in, xii, 74, 75, 76, 198, 203
sex blindness in, xii, xiv, xvi, xix, 76, 198
Equal Protections Clause, 168–69, 170, 172, 173, 184, 209
as tool to dismantle sex discrimination, 174–76, 177, 178, 185, 189
Equal Rights Amendment (ERA), 171
estrogen, 13, 14, 17, 18, 22, 70, 93, 121, 122
European Court of Human Rights, 33
European Heart Journal, 130
European Union, 109
Evangelista, Angel, 210
Evangelista, Elektra, 210
Extraordinary Congress, 114

Fadel, Hossam, 93
Faludi, Susan, 156
father, as defined in law, 46–47
Federalist Society, 225
Felix, Allyson, xvii
female athletes, 61, 109–10, 243
in art, 108, 263
biological differences as overlooked in training of, 117–18
misogynoir and, 28
puberty in, 116, 117–18
see also female sports
female body, 10, 43, 87, 93, 130, 137, 140, 243, 251, 259
abortion rights and, *see* abortion, abortion rights
art in celebration of, 107–8, 109, 110, 241, 263–64
beauty in, 106, 107–8, 118, 232, 263–64
as initially ignored in medical research, 121–22

puberty and changes to, 9, 13–14, 17–18, 48, 116, 117, 154, 155, 250
sex discrimination as based on, xviii
sexualization of, 111–12
sexual response cycle in, 99
in utero development and, 12–13, 21
see also females; sex characteristics; sex differences; sexed bodies
female puberty, 16, 21, 116–18, 133
in athletes, 116, 117–18
body changes due to, 9, 13–14, 17–18, 48, 116, 117, 154, 155, 250
dropping age in onset of, 154–55, 248
mental health issues and, 21, 48, 142, 151–56, 218, 249
females:
abortion rights and, *see* abortion, abortion rights
anxiety and depression in, 21, 48, 128, 151–52, 155, 218, 249
as barred from certain professions, 34–40, 52
black, *see* black women
body and form of, *see* female body
caregiving roles of, 147–49, 150, 151
conservatism and "lack of agency" in, 236–37
cultural artifacts attached to, 50–52, 53
discrimination against, *see* sex discrimination
domestic violence and, 142, 143, 144, 145, 146, 226
as having fewer children, 90, 91–92
hormone fluctuations in, 21–22, 28, 116, 117
late-onset gender dysphoria in, 248–49
law's definition and treatment of, 34–35, 37–41, 52, 53, 56, 57, 137–38, 141, 169, 170, 171, 172, 174, 176–77, 178, 179, 192; *see also* sex classifications
in military, 139–40, 178–79
policies and norms as disproportionately harmful to, 140–42, 145–46, 147, 148–50, 155
puberty in, *see* female puberty
redefining of term, 77, 79, 256, 259, 260
"running while," 144
sex gaps and, *see* sex gaps
sexual violence against, *see* sexual violence
sidelining and silencing of, in progressive advocacy, xv, 62, 222–24, 225–27, 245–46, 251–52, 260–61, 264
traditional roles of, 37–38, 39–40, 89, 110, 147, 148

in Ukraine invasion, 138–39, 140
see also biological sex; sex differences
female sports, xii, xiii, 216
benefits of, xiv, xvii, 118–19, 201, 202
in Equality Act, xiii, xvi–xvii, 200–202, 204
sex-based eligibility standards and, xiv, 7, 28, 29–33, 60–62, 114–15, 119–21, 124, 205, 238–39, 250, 252–54
sex segregation as beneficial for, xiii, xiv, 27, 62, 115–16, 119–20, 202, 251, 252, 253
Title IX and, xiii, 200–202, 204, 206, 225
transgender women and, xx, 60, 61–62, 69, 77, 79, 120, 201, 204, 205, 225, 245, 251–52, 253, 259–60
femininity, 67, 69, 207
beauty in cultural expressions of, 241–42
gender identity and, 74
social construction of, 50–51, 69
feminism, feminists, xx–xxi, 35–36, 63–64, 110, 187, 232, 261, 262, 263
radical left's attack on opponents to sex blindness in, 223–27
sex-blind identity faction of, xviii, 28–29, 64–65, 67–68, 74, 76, 87, 203, 215
sex-linked biology faction of, 64–65, 66–67, 68, 87, 101, 176–77, 214–15, 222, 223–25, 226, 227, 261
transgender rights in, xx–xxi, 64, 65, 66–67, 76–77, 203
see also progressive advocacy; women's rights; women's rights movement
fetus development, 11, 12, 24
FINA (Fédération Internationale de Natation; now World Aquatics), 114–15
First Amendment, 225
5-alpha reductase deficiency (5-ARD), 24, 70
Florence, Italy, 85, 112
Florida, 72, 111
"Don't Say Gay" law in, 217–18, 219
food stamps, 149
Fourteenth Amendment, 36, 37, 53, 167–69, 176, 184, 185
Due Process Clause of, 168, 209
Equal Protections Clause of, *see* Equal Protections Clause
Privileges or Immunities Clause of, 37, 168
Fox News, 225
France, 107, 234
Frankfurter, Felix, 171–72
Free speech, 220, 221
Friedman, Jane M., 54

Frontiero, Sharron, 167, 178-79, 180
Frontiero v. Richardson, 178-80, 236

Gage, Frances Dana Parker, 277n
Gaines, Riley, 205
Galleria Dell'Accademia (Florence), 85, 111
gay rights, xiv, xv, 64, 65, 196, 202, 203, 211, 221
    Equality Act and, xi, xii, xvii, 74, 75, 76, 77,
      199, 200, 230
    landmark cases in, xxi, 209, 210-11, 236
    political backlash to, 217-18
    radical left's hijacking of movement for,
      215-16
"Gay Rights and the Limits of Liberalism"
    (Sullivan), 215-16
gender:
    beauty in cultural expressions of, 241-42
    definition of, 8, 239
    religious right's view of sex and, 206-7
    sex as conflated with, 44, 45, 79-80, 131,
      180, 189, 190-91, 211, 215, 229, 239,
      240, 255
    sex vs., 8, 25-26, 29, 33, 40-41, 45, 47, 65,
      120, 123, 131, 239-41, 242, 253
gender-affirming care, 16, 17, 18, 26, 32, 43,
    65, 66, 207, 231, 253, 256, 257, 259-60
    common sense approach to, 246-50
    conservative backlash to, xx, 217, 218-19,
      249
    distinguishing transgender people based on,
      256, 258-59
    prosecuting of parents for, 218-19
    puberty blockers in, 18, 67, 71, 216, 246,
      248, 249-50
    for trans kids, xx, xxi, 18, 71, 216, 217,
      218-19, 246-50
gender-bending expressions, 241-42
gender diversity, gender-diverse people, xix, 23,
    65-68, 120, 121, 124, 239
    brain development and, 25, 70
    civil rights for, 65-67, 208, 209 *see also*
      transgender rights
    normalizing of, 210
    see also disorders of sex development (DSD);
      intersex people; nonbinary identities;
      specific transgender headings
gender dysphoria, 94-95, 197, 208, 218, 240,
    247, 249, 259
    early-onset, 25, 247
    late-onset, 248-49
    see also specific transgender entries

gender expression, 6, 64, 73, 206, 208, 248,
    254
    beauty in, 241-42
gender identity, 66, 94, 97, 101, 102, 142, 197,
    199, 206, 212, 214, 217, 227, 231, 242,
    244, 251
    biological sex vs., 25-26, 29, 33, 40-41, 45,
      47, 65, 123, 131, 239-40, 242, 252, 253
    brain and, 25, 70, 71-72, 247
    discrimination based on, xi, xii, xxi, 199, 209
    preferred pronouns as expression of, 254-55
    redefining sex as, xii, 29, 44, 64, 70-72, 75,
      79-80, 211, 256
    sex-blind faction of feminism as focused on,
      28-29, 63, 64, 65, 67-68, 74, 76-77, 87,
      203, 215
    viewed as primary sex characteristic, 70,
      71-72, 207
genes, 21, 23, 128, 130
genitals, 9, 25, 69, 247
Georgia, 57, 104
Ginsburg, Ruth Bader, 45, 53, 64, 167, 171,
    174, 175-77, 180-81, 187, 189, 196, 198,
    200
    Frontiero v. Richardson case and, 178-80
    Reed v. Reed brief of, 177-78, 179
    sex differences as viewed by, xix, 46, 87,
      193-94, 195
    VMI opinion of, xix, xx, 46, 192-94, 195
Goldberg, Michelle, 64-65, 210, 214-15, 216
Goodell, Lavinia, 36
Graham, Robert, 108, 263
Grinberg, Charlotte, 96
Grinberg, Raffi, 96-97
Griswold v. Connecticut, 137
groupthink, 221
Gu, Eileen, 110
Guardian, 5, 263
Guma, Elisa, 125

Haines, Julia, 147
Hambling, Maggi, 263
Hank (author's father), 232, 233, 235
"happiness penalty," 97
Harinam, Vincent, 158
Harle, Denise, 236-37
Harper, Joanna, 259-60
Harvard Business Review, 50
Harvard Law School, 35, 58, 171
Harvard-Radcliffe College, 58
Harvard University, 19, 49, 245

Harvey, Marceline (Marcelin Harvey), 212–13, 260
heart disease, 129-30, 155, 161
hermaphroditism, 24, 26
  see also ovotesticular disorder
heterosexuality, 20, 99, 102, 103, 206
  see also sexual orientation
High Court of the United Kingdom, Family Division of, 46-47
Hill, 161
Hill, Brianna, 52-53
Hiltz, Nikki, 253, 254
Hinduism, 104
Hirschberg, Angelica, 21
Hmong refugees, 229
Hogshead, Nancy, 225
Hollberg, Cecilie, 111
homosexuality, 20, 46, 99, 103-4, 143, 209, 241, 248
  conversion therapy for, 209
  see also gay rights; LGBTQ communities; sexual orientation
Hooven, Carole, 19
hormones, 93-94, 95, 99, 116, 117
hormones, sex, 8, 9, 12, 13, 14, 15, 25, 28, 93, 94, 120, 121, 129, 135
  as causing sex differences in brain, 19, 128, 130, 152, 154
  in disorders of sex development (DSD), 23, 24, 29, 70, 238
  in gender-affirming care, 17, 18, 26, 66, 67, 71, 73, 231, 246-47, 248, 249-50, 253, 259-60
  post-puberty and beyond, 21-22, 130
  see also estrogen; progesterone; testosterone
House of Representatives, U.S., 198, 200
  Judiciary Committee of, Equality Act hearings and, xi, xiv-xix, 61, 230
Hulett, Alta, 54
Human Rights Campaign (HRC), 69, 212, 213, 215
Hunt, Taylor, 108
Hunter, Sandra, 253

Idaho, 177
Illinois Supreme Court, 34-35, 36, 37, 38, 54
Independent Council on Women's Sports, 245
Indiana, 138
Indian Health Service, 162
Indian reservations, 143, 162
India Times, 50

infant formula, 51
Instagram, 153
Institute for Family Studies, 91
Institute of Medicine (IOM; now National Academy of Medicine), 8, 9, 40, 122, 123, 131
intermediate scrutiny, 181, 182, 185-86, 194
International Amateur Athletic Federation (now called World Athletics), 30
International Monetary Fund (IMF), 148
Intersex Human Rights Australia, 25
intersex people, intersex traits, xii, 8, 9, 20, 22, 23, 32, 61, 62, 75, 76, 77, 214, 238
  see also disorders of sex development (DSD)
Invisible Women (Criado Perez), 141, 144, 148
Iran, 41
Iranian Revolutionary Guard, 146
Islamic traditions, 93
Ivy, Veronica (Rachel McKinnon), 100

Jackson, Ketanji Brown, 55-56, 57-58, 80-82
  confirmation hearings of, 55-56, 57, 58-60, 62-63, 76-77, 78, 79
Jackson, Leila, 58, 59
Jackson, Talia, 58
Jane Crow (Rosenberg), 174
Japanese Americans, 172, 199
J.E.B. v. Alabama, 189-91, 195, 196
Jefferson, Thomas, 258
Jenner, Caitlyn, xx, 25-26, 253, 260
Jim Crow, 58, 168, 257
John, Elton, 215
Johns Hopkins University, 229
Johnstone, Faye, 223
Jones, Kim, 245
Joyner, Mike, 252
Justice Department, U.S., 142

Kama Sutra, 104
Kardashian, Kim, 108
Karolinska Institute, 21
Kecmanovic, Jelena, 152
Kennedy, Anthony, 210
Kennedy, John F., 174
King, Martin Luther, Jr., 81, 232, 233
Klein, Ezra, 210
Klein, Sabra, 132
Kois, Dan, 111
Korematsu v. United States, 172, 173
Kratochvilová, Jarmila, 4

Lansky, Greg, 108
Larsen, Nella, 257
Lasitskene, Mariya, 110
Lausanne, Switzerland, 109, 133–34, 232
law, legal system:
  black women as treated in, 56, 57
  children of enslaved people in, 56–57
  failure to ratify ERA into, 171
  "father"/"mother" as defined in, 46–47
  females as defined and perceived in, 34–35,
    37–40, 52, 53, 56, 57, 137–38, 141,
    169, 170, 171, 172, 174, 176–77, 178,
    179, 192; *see also* sex classifications; sex
    discrimination, sexism
  gay rights landmark cases in, xxi, 209,
    210–11, 236
  intermediate scrutiny rule in, 181, 182,
    185–86, 194
  persecution of gays and gender-diverse people
    in, 208–9
  police power in, 42, 179
  racial discrimination cases in, xvi, 172–74
  rational basis test in, 175, 177–78, 180
  sex as defined in, 41, 43–44, 45, 52, 53
  sex classifications in, *see* sex classifications
  sex differences characterized as "myths" or
    "stereotypes" in, xix, 91, 189–90, 191, 199
  sex discrimination landmark cases in, xix,
    167, 176, 177–80, 181–83, 185–96
  sexual violence as propped up by, 145–46
  strict scrutiny test in, 172–73, 177, 180, 181,
    186, 187, 188, 191, 192, 193, 194
  two roles of, 41–42
  see also specific court cases
Lawrence v. Texas, 209
Ledecky, Katie, 252
lesbians, 99, 103, 222, 227
  Johns Hopkins definition of, 229–30
  transgender women as, 101
Levy, Ariel, 4, 5–6, 32
LGBTQ communities, xi, xiv–xv, xvi, xvii, xix,
    xx–xxi, 64–65, 76, 196, 197, 202, 218,
    220, 229–30
  banning of school instruction on issues of,
    xx, 217–18, 219
  civil rights for, *see* gay rights; transgender
    rights
  desire to procreate in, 97–98
  high-risk kids in, 218, 241, 249
  mainstream normalizing of, 210
  see also transgender people

Lily (author's grandmother), 234
Lincoln, Abraham, 54
Lincoln, Mary Todd, 54
Lincoln, Robert, 54
Liu, Siyuan, 125
locker rooms, 245
London, England, 263
loneliness, 161–62
Longman, Jere, 32
Looking Down the Avenue (Dine), 107
Lopiano, Donna, 225

Mahuchikh, Yaroslava, 110
male body, 10, 43, 93, 106, 130, 140, 242, 243
  art in celebration of, 85–86, 107, 108, 109,
    110–11, 112–13, 241, 263
  beauty in, 85, 86, 106–7, 108, 110, 113,
    140, 241
  puberty and changes to, xvi–xvii, 9, 13,
    15–17, 32, 49, 116, 238, 250–51
  sexual response cycle in, 99
  in utero development and, 11–12, 24
  see also males; sex characteristics; sex
    differences; sexed bodies
"male malaise," 156–57
male puberty, 13, 15–17, 20, 49, 116, 154
  body changes due to, xvi–xvii, 9, 13, 15–17,
    32, 49, 116, 238, 250–51
  sex-based eligibility standards and advantage
    of, xvi–xvii, 29, 31, 32, 115, 238, 251, 252,
    253, 259–60
  testosterone exposure and, xvi–xvii, 32, 69
males:
  aggression in, 20, 43, 49, 127, 144
  athletic performance advantage of, xvi–xvii,
    27, 31, 48, 49, 115, 120, 250–51, 252,
    259–60, 261–62
  benefits of procreation for, 94
  body and form of, *see* male body
  caregiving roles and, 147, 148
  cultural artifacts attached to, 50–52, 53
  as defined in law, 41
  domestic violence and, 143
  early onset gender dysphoria in, 25
  educational underperformance of, 156–58,
    161
  gender gaps and, *see* gender gaps
  hormones in midlife and beyond for, 22
  military drafts and, 44–45, 136, 139, 141,
    162, 163
  puberty in, *see* male puberty

sexual violence and, 87, 142–46, 162, 235–36, 243
social isolation in, 161–62
suicide and, 21, 162
traditional roles of, 37–38, 39–40, 89
workplace participation as dropping among, 158–61
see also biological sex; sex differences
Maney, Sarahbeth, 58, 59
Manuel, Simone, xvii
Marcelin, Harvey (Marceline Harvey), 212–13, 260
Maria (mixed-race slave), 56
Marrakesh, 93
Marshall, Thurgood, 55, 81, 173, 180, 189
"Mary on the Green" group, 263
masculinity, 67, 69, 87
beauty in cultural expressions of, 241–42
gender identity and, 74
social construction of, 50–51, 69
toxic, 87, 102, 146, 236
Master Slave Husband Wife (Woo), 169, 257
Material Girls (Stock), 99, 101, 227
matriarchal societies, 147–48, 237
Mayo Clinic, 129
McCain, John, xiv
McCarthy, Margaret M., 128
McConnell, Freddy, 26, 47, 98
medical research, consideration of sex differences in, 45, 121–26, 131
Medium, 203
Medoff, Carmi, 150–51
melanoma, 133–35
menarche, 14, 91, 154, 155
meningioma, 129
menopause, 14, 18, 22, 92, 95, 130
menstrual cycle, 13, 17, 18, 21–22, 48, 70–71, 93, 116–17, 118, 122
mental health:
female adolescence and, 48, 142, 151–56, 218, 249
sex differences in, 21, 48, 128, 141, 151–56, 161–62, 218, 249
sexual and domestic violence effect on, 145
social isolation and, 161–62
social media and, 152–53
Men Without Work: America's Invisible Crisis (Eberstadt), 158
merger doctrine, 38, 169
Mernissi, Fatima, 261
Michelangelo, 85–86, 107, 111, 112, 113

migraines, 21, 48, 128
military:
gays and transgender people in, 209
women in, 139–40, 178–80
military drafts, 44–45, 136, 139, 141, 162, 163, 232
Miller, Virginia, 129–30, 131
mini-puberty, 15, 25, 247
misogynoir, 27–28
misogyny, xix, 62, 64, 76, 101, 124
Mississippi, 146, 147, 185, 186
Mississippi University for Women v. Hogan, 185–87, 188, 193
mixed race children, in slavery, 56–57
Modern Family (TV series), 97, 210
modesty, 111–12, 263
mood-related disorders, 21, 48
mothers:
as defined in law, 46–47
single, 148–49, 233–34
Motley, Constance Baker, 81
Mulvaney, Dylan, 213, 219, 292n–93n
Mumbai, 143
Munro, Neil, 203–4
Murray, Pauli, 173–76, 177, 179, 192, 196–97
Murthy, Vivek, 153
Muslims, 44, 93
myelination, 126, 127
My Name Is Pauli Murray (documentary), 174

Nadal, Rafael, 53
Nation, 37, 49
National Academy of Medicine, *see* Institute of Medicine
National Center for Health Statistics, 162
National Institutes of Health (NIH), 45, 122, 123
National Park Service, 234
National Women's Law Center (NWLC), xvii–xix, 61–62, 64, 76–77, 203
Navratilova, Martina, 200–201, 202–3, 204, 225, 229–30, 252
NBC News, 61, 229
NCAA National Championships (2022), 205
New World Encyclopedia, 105–6
New Yorker, 4–5, 6, 154
New York State Woman Suffrage Association, 170
New York Times, 29, 32, 58, 64, 74, 91, 132, 153, 210, 212, 214, 224–25
Nikolai (Russian draftee), 162–63

Nineteenth Amendment, 170
nonbinary identities, 25, 65, 67, 70, 73, 74,
    204, 208, 210, 211-12, 213, 241, 244, 253
North Carolina, 163-64, 244
North Carolina, University of, at Chapel Hill, 196

Obama, Barack, 211
Obergefell v. Hodges, 209, 210-11, 236
O'Connor, Sandra Day, 187, 189-90, 191, 192,
    193, 195, 196, 200
Of Boys and Men (Reeves), 156-59
Ohio, 138
Olson, Ted, 194
Olympic Games, xvii, 3, 6, 109, 110, 114, 116,
    121, 133, 205
    of 1976 (Montreal), 25, 253
    of 2016 (Rio de Janeiro), 3, 29, 30, 31, 114,
    133
Olympic Gateway (Graham), 108, 263
Olympic Museum (Lausanne), 109
Olympic Village, 109
"one-drop rule," 258
Onsite Kids, 150
orgasm, in sexual response cycle, 99
Ostrom, Quinn, 129
ovaries, 9, 10, 11, 12, 13-14, 70, 231
ovotesticular disorder, 9, 24, 26, 214
Oxford Dictionaries, xii, 75
Oxford English Dictionary, 7
Oxford Languages, 207

Paoletti, John, 85, 108, 112-13
"Parental Rights in Education" law, 217-18,
    219
Paris, France, 107, 109, 114
partus sequitur ventrem, 56, 57
"passing," 257-58
Passing (Larsen), 257
"Pass the Equality Act but Save Title IX"
    (*Washington Post* op-ed), 200-201, 203, 225
patriarchy, patriarchal societies, 50, 51, 52, 53,
    54, 65, 68, 118, 124, 147-48, 156, 222,
    236, 262
pediatric trans medicine, 246-48, 249
Pennsylvania, University of, 60, 245
peremptory challenges, 189-91
performance gaps, xvii, 16, 31, 49, 115, 120,
    250-51, 252, 259-60, 262
Persky, Aaron, 145-46
Personnel Administrator of Massachusetts v. Feeney,
    184-85

Phelps, Michael, 116
Piazza della Signoria (Florence), 112
Poland, 90, 138
population, 88, 89
    declining numbers in, 90-92
Portugal, 90, 97
Pose (TV series), 210
post-traumatic stress disorder (PTSD), 145
poverty, 140-41, 149
Powell, Jeff and Sarah, 197
Powell, Lewis, 180
pregnancy, xii, 17, 18, 21, 64, 76, 91, 92,
    93-94, 133, 137, 171, 226, 232
    in transgender men, 26, 98, 228
    see also abortion, abortion rights
primary sex characteristics, 11-14, 15, 69-71,
    75, 228
    gender identity viewed as, 70, 71-72, 207
    in "sex is social construct" argument, 70-72
Privileges or Immunities Clause, 37, 168
procreation, 87, 88-98, 106, 214
    declining numbers in, 90-92
    individual benefits of, 92-98
    in religious communities, 93, 95-96
    societal benefits of, 88-92
    see also reproduction, sexual
Procter & Gamble, 212
progesterone, 14, 93, 121, 122
progressive advocacy, 204, 209, 211-19,
    243-44
    assault on free expression in, 219-30
    females as sidelined by, xv, 62, 222-24,
    251-52, 260-61, 264
    opportunistic use of black women in,
    60-61
    overlapping goals with other social justice
    groups in, 213
    political backlash to, 216-19
    preferred pronouns pushed by, 255
    private sector targeted in, 211-13
    redefining of sex as strategy in, xii, 44, 63,
    64-65, 68-80, 198, 203, 207, 211, 213,
    215, 231
    sex-based eligibility standards opposed by, 28,
    29-30, 32, 60-62, 124
    sex-blind strategy of, *see* sex blindness
    sex differences viewed as "myths" and
    "stereotypes" in, xviii-xix, 69, 70, 75, 199,
    200, 203, 205
    sex viewed as "social construct" in, 68-72,
    199, 207

silencing of dissenting views by, xv, xviii, 223–27, 241, 249, 260, 264
"trans women are women" stance in, xviii, 60, 76, 102, 203, 227, 260
see also feminism, feminists
pronouns, 67, 74, 207, 212, 253
common sense approach to, 254–55
property rights, 54, 169
Pryor, Jill, 71–72
puberty, xvi, 20, 25, 29, 31, 32, 67, 115, 260
age of onset of, 154, 155, 248
development of secondary sex characteristics in, 9, 13–14, 15–18, 48, 49, 69, 116–18, 154, 155, 250
gender-affirming treatment at onset of, 247–48
see also female puberty; male puberty
puberty blockers, 18, 67, 71, 155, 216, 246, 248, 249–50
Putin, Vladimir, 163

Queen's Gambit, The (TV series), 127
Quillette, 158, 201
Quran, 93, 95

race-sex analogy, 175, 177, 179, 188, 192–93, 195, 196
racism, racial discrimination, 57, 59–60, 62, 172, 174, 175, 179, 184, 188, 196
"strict scrutiny" test and, 172–73, 187, 191
structural, 57, 125, 154
see also segregation, racial
radical left, 207–8, 213, 226
see also progressive advocacy
rape, 56, 57, 137, 138, 140, 142, 143, 144–45, 189, 229
see also sexual violence
Rapinoe, Megan, xx, 110
Raseboya, Violet, 5, 27, 242
rational/reasonable basis test, 175, 177–78, 180
Raznahan, Armin, 125
Reconstruction, 168
Reed, Sally, 176, 177, 178
Reed v. Reed, 176, 177–78, 179, 180, 181, 193
Reeves, Richard V., 156–59
Rehnquist, William, 182, 190
"Reimagining Venus de Milo in the Modern Age" (Hunt), 108
religious right, 206–8, 216, 249
see also Republican Party, Republicans
replacement rate fertility, 90, 91, 92, 149

reproduction, sexual, 41, 42, 46, 72, 125, 262
sex characteristics in facilitating of, 10, 12, 13, 15, 19, 22, 70, 93, 238, 256
see also procreation
Republican Party, Republicans, xii, xiv, xv, 200, 204
in political backlash against trans rights, xix–xx, 60, 216–19
Rhoades, Everett, 162
Rhode Island, 147
Richards-Ross, Sanya, xvii, 200–201, 202
Rodin, Auguste, 107, 108
Rodriguez, Bianca, 210
Roe v. Wade, 137, 168
Roosevelt, Eleanor, 174
Rosenberg, Rosalind, 174
Rosin, Hanna, 156
Rowling, J. K., 102, 226
"running while female," 144
Russia, 109–10
military conscriptions in, 162, 163
Ukraine invaded by, 136, 138–40, 162–63

St. James, Emily, 218
Sako, Phineas, 4, 5–6
same-sex marriage, 209, 210–11, 215–16
Sami people, 228–29
Scalia, Antonin, 189, 190–91, 194–95, 196
Schatz, Howard, 109
scientific research, consideration of sex differences in, 45, 121–26, 131
Sculpture for Mary Wollstonecraft, A (Hambling), 263
Seahorse: The Dad Who Gave Birth (documentary), 26, 47, 98
secondary sex characteristics, 9, 15–18, 75, 116–18, 154, 155, 175, 228, 250
in "sex is social construct" argument, 68–69
see also female puberty; male puberty; puberty
Secretariat, 252
segregation, racial, 173
segregation, sex:
in education, 194
in public spaces, 242–46
in sports, xiii, xiv, 27, 62, 115–16, 119–20, 202, 250–51, 252, 253
Sekgala, Maphuti, 5
Selective Service System, 44
Seme, Michael, 5

Semenya, Caster, xiv, 3-7, 27-28, 29, 30-31, 32-33, 64, 206, 224, 238-39, 242, 253
Semenya, Jacob, 6
Senate, U.S., xiv, 58, 80, 200
    Jackson's confirmation hearings in, 55-56, 57, 58-60, 62-63, 76-77, 78, 79
"separate but equal" doctrine, 58, 168, 173
sex:
    as biologically defined, *see* biological sex
    cultural artifacts and, 50-52, 53
    as defined in law, 41, 43-44, 45, 47, 52, 53
    as defined in progressive advocacy groups, *see* progressive advocacy
    differences based on, *see* sex differences
    disconnecting sexual orientation from, 102-4
    discrimination based on, *see* sex discrimination, sexism
    gender as conflated with, 44, 45, 79-80, 131, 180, 189, 190-91, 211, 215, 229, 239, 240, 255
    gender identity viewed as, xii, 29, 44, 64, 70-72, 75, 79-80, 198, 211, 256
    gender vs., 8, 25-26, 29, 33, 40-41, 45, 47, 65, 120, 123, 131, 239-41, 242, 253
    "nonbinary" view of, 11, 70-71, 207, 215, 239, 244; *see also* sex blindness
    procreation and, 88-98
    progressive advocates in redefining of, xii, 44, 63, 64, 68-80, 198, 203, 207, 211, 213, 215, 231; *see also* sex blindness
    race analogy to, 175, 177, 179, 188, 192-93, 195, 196
    religious right's view of gender and, 206-7
    "social construct" view of, 68-72, 199, 207
    see also females; males; sex classifications
sex-based eligibility standards, xiv, 7, 28, 29-33, 114-15, 119-21, 124, 205, 238-39, 250
    black women in debate on, 60-62
    common sense approach to, 250-54
    male puberty advantage and, xvi-xvii, 29, 31, 32, 69, 115, 238, 251, 252, 253, 259-60
sex-based spaces, 27, 66, 242-46
    see also bathrooms
sex blindness, xiii, xiv, 58, 60, 192, 196, 221
    attacking and silencing of opponents of, xv, xviii, 223-27, 241, 249, 260, 264
    changing language and definitions in pursuit of, xii, xxi, 29, 44, 63, 64-65, 67, 68, 72-74, 76, 77-80, 198, 203, 227-30, 240, 259

conflation of sex and gender in, 44, 45, 80, 131, 180, 189, 190-91, 211, 215, 229, 239, 240, 255
    and dismissal of sex differences as "myths" or "stereotypes," xviii-xix, 69, 70, 72, 75, 124, 183, 188, 189-90, 191, 194, 199, 200, 203
    in Equality Act, xii, xiv, xvi, xix, 76, 198
    erasure of biological sex in, xii, xv, xviii, xix, 29, 31, 44, 63, 64-65, 67-68, 70-71, 72, 73, 75-76, 87, 198, 199, 203, 207, 211, 213, 214-16, 222, 227, 240, 259
    in feminist movement, xviii, 28-29, 64, 67-68, 74, 76, 87, 203, 215
    Ginsburg's rejection of, xix, 46, 87, 194, 195
    in private sector, 211-13
    pronouns in advancing of, 255
    sex smart vs., xv, xix, xx, 162, 240
    see also progressive advocacy
sex characteristics, xii, 8, 9-14, 15, 21-22, 26-27, 29, 75, 76, 121, 198
    binary sorting of, 9-10, 11, 14, 22, 24-25, 69-70, 103, 238
    exceptions to binary sorting of, 22-26, 29-33, 70, 238
    in facilitating reproduction, 10, 12, 13, 15, 19, 22, 70, 93, 238, 256
    in intersex/DSD people, 9, 23, 24-25, 238, 239, 257
    mini-puberty and, 15, 25
    primary, *see* primary sex characteristics
    puberty and development of, 9, 13-14, 15-18, 48, 49, 69, 116-18, 154, 155, 250
    secondary, *see* secondary sex characteristics
    in "sex is social construct" argument, 68-72, 207
    in utero development of, 10-14, 15, 20, 21, 23, 24, 25
sex classifications, xii-xiii, 42, 43-44, 66, 123, 170-71, 184
    Ginsburg on, 194, 195
    intermediate scrutiny of, 181, 182, 185-86, 194
    legal challenges to, 174-75, 176, 177-78, 179, 180-83, 185-87, 188-96; *see also specific cases*
    military drafts and, 44-45
    progressive advocacy's goal for dismantling of, 67-68, 76; *see also* sex blindness
    rational basis test for, 175, 177-78

strict scrutiny of, 177, 180, 181, 186, 188, 192, 193, 194
see also biological sex; sex discrimination
sex differences, xviii, 21–22, 42, 65, 140, 148, 188, 194
in athletic performance, xvi–xvii, 27, 31, 48, 49, 115, 120, 250–51, 252, 253, 259–60, 261–62
in behavior, 20–21, 123, 127, 188
in brain and brain development, 19–20, 21, 48, 49–50, 125, 126–29, 130, 152, 154–55, 157, 248
in cardiovascular system, 129–31
in disease patterns, 43, 123, 128, 129, 131, 132, 134–35, 161
educational performance and, 156–58
Ginsburg's view of, xix, 45–46, 87, 193–94, 195
in head shape and facial features, 16, 17–18, 26–27
in immune system, 131–32, 134–35
importance of recognizing, 189–90, 193–94, 195–96, 230, 231
medical/scientific research and consideration of, 45, 121–26, 131, 240
in mental health issues, 21, 48, 128, 141, 151–56, 161–62, 218, 249
as "myths" or "stereotypes," xviii–xix, 69, 70, 72, 75, 124, 183, 188, 189–90, 191, 194, 199, 200, 203, 205
as overlooked in sports training, 117–18
reasonable vs. unreasonable inferences based on, 47–50, 52, 53, 124
in response to social media, 152–53
sex-based public spaces and, 243
in skin, 133–35
see also gender gaps; sex characteristics
"Sex Differences in the Human Brain" (DeCasien et al.), 125, 127, 141
sex discrimination, sexism, xv, xvi, xx, 45, 53, 64, 65, 74, 87, 125, 127, 139, 141, 174, 198, 199, 214, 243
as based on female biology, xviii, 64
Equal Protections Clause as remedy for, 174–76, 177, 178, 185, 189
"exceedingly persuasive justification" in cases of, 184–86, 193, 194
jury peremptory challenges and, 189–91
landmark cases in, xix, 167, 176, 177–80, 181–83, 185–96; *see also specific cases*

race analogy to, 175, 177, 179, 188, 192–93, 195, 196
sex classifications and, *see* sex classifications
in workplace and employment, 34–40, 44, 54, 171–72, 183
see also women's rights
sexed bodies, 14, 67, 196
beauty in, 85, 86, 106–13, 140, 232, 241, 263–64
calls for modesty and, 111–12
procreation and, 88–89, 90, 91–92, 93–95
sexual pleasure and, 98–105, 113, 140
see also biological sex; female body; male body; sex characteristics;
sex equality, 170, 174, 175, 177, 179, 182, 187, 190, 196, 214
in elite sports, 115–16, 119–20
ERA and, 171
as not goal of early women's rights movement, 169–70
see also sex discrimination, sexism; women's rights
sex gaps:
in athletic performance, *see performanc gaps*
in child care impact, 147, 148–50
in educational performance, 156–58, 161
in workplace participation, 158–61
Sex in Law, 226
"Sex in Sport" (Coleman), xiv
sexism, *see* sex discrimination, sexism
"Sex Neutrality" (Coleman), 226
sex segregation, in sports, xiii, xiv, 27, 62, 115–16, 119–20
sex-smart, 121, 200
sex blindness vs., xv, xix, xx, 162, 240
sexual assault, 112, 142, 143–46, 162, 240, 241
see also rape; sexual violence
sexual orientation, xii, 19, 74–75, 97, 99, 105, 142, 196, 197, 198, 217
disconnecting sex from, 102–4
discrimination based on, xi, xii, xxi, 66, 199, 208–9
as "transphobic," 100–102
see also gay rights; homosexuality
sexual pleasure, 87, 98–105, 106, 113, 140
mental and physical benefits of, 104
privacy rights and, 104–5
sexual relations, 98–99, 102, 106, 111, 137, 155, 241
privacy rights and, 104–5
see also sexual pleasure

sexual response cycle, 98–99
sexual violence, 64, 87, 112, 137, 138, 142–46, 162, 243
  collective harm of, 145
  common sense approach to, 235–36
  as propped up by laws and social norms, 145–46
  in slavery, 56
  in wartime, 139–40
  see also rape; sexual assault
Shah, Nirao, 27
Shappley, Kai, 25
Shelley, Mary, 261
Siegel, Neil, 192
single parents, 140, 147, 148–49, 233–34
skin cancer, 133–35
skin-fold tests, 116, 117
Slate, 111
slave codes, 168
slavery, 56–57, 58, 59, 93, 167, 169, 221, 230, 257
  mixed-race children in, 56–57
  women in, 51, 56, 57, 59
Smith, Cornelia, 196
Smith, Geraldine and Stephen, 234
Smith, James, 56, 57, 169
Smith, Mary Riffin, 197
Smith, Victoria, 68
Smithsonian Institution, 208
soccer, xvii, 4, 5, 6, 110
social isolation, 161–62
social media, 21, 53, 101, 110, 203, 263
  mental health impact of, 152–53
South Africa, 33
South Carolina, 147, 164
Southern Poverty Law Center, 179
sperm, 10, 11, 12, 13, 15, 16, 17, 91, 93
  donation of, 92
sports, xiii, 201, 241
  anti-doping programs in, xiii–xiv
  art's celebration of athletic body in, 108–10
  benefits of sex segregation in, xiii, xiv, 27, 62, 115–16, 119–20, 202, 251, 253
  elite, *see* elite sports
  female biology as overlooked in training for, 117–18
  investment gap between men's and women's, 206
  male performance advantage in, xvi–xvii, 27, 31, 115, 120, 250–51, 252, 259–60

youth, community, and amateur, 115, 119, 121, 250–52
  see also female sports; *specific sports*
Sports Illustrated, 26, 110
SRY gene, 11–12
Stanford University, 110, 145, 187
State Department, U.S., Sports Diplomacy division of, 110
States' Laws on Race and Color (Murray), 173, 175
Stewart, Potter, 180
Stock, Kathleen, 99, 101, 226–27, 261
Stone, Lyman, 91
Strangio, Chase, 61, 203
Strickland, Patrick, 149
strict scrutiny test, 172–73, 177, 180, 181, 186, 187, 188, 191, 192, 193, 194
structural racism, 57, 125, 154
suicide, 21, 145, 151, 152, 159, 162, 218, 233, 240–41, 249
Sullivan, Andrew, 102–4, 113, 215–16
Supreme Court, U.S., 54, 58, 79, 81, 124, 170, 171–72, 176, 187, 199, 200
  changes in composition of, 188–89
  Jackson's confirmation hearings for, 55–56, 57, 58–60, 62–63, 76–77, 78, 79
  see also specific Supreme Court cases
Supreme Court of India, 105
surgery, gender-affirming, 26, 66, 71, 246, 247
surrogacy, 92, 140
Survey Center on American Life, 161
swimming, xvii, 60, 114, 116, 120, 121, 205, 245, 252
  skin-fold tests in, 116, 117
Swiss Federal Tribunal, 33
Switzerland, 33, 61, 109, 232, 234

Tallungs, Kaj, 88
Team USA, 250, 253
Telegraph, 3
telenovelas, 146
Tennessee, 170
TERFs ("trans exclusionary radical feminists"), 64, 101, 222–23, 224, 225
testicles, 9, 10, 11, 12, 13, 16, 17, 69, 70, 77, 238
testosterone, xvi, 12, 13, 15, 26, 94, 98, 121, 224, 231
  aging and decline in, 22
  CAIS and, 24
  male puberty and, xvi–xvii, 29, 32, 69, 93

sex-based eligibility standards and, xvi–xvii,
29, 30, 32, 60, 238–39, 253
*Testosterone: The Story of the Hormone That
Dominates and Divides Us* (Hooven), 19
Texas, 50
prosecution of gender-affirming care, 218–19
thelarche, 154
Thomas, Clarence, 55, 56, 189, 190
Thomas, Lia, 60, 62, 205, 245, 253, 259–60
Thomas, Peyton, 74
TikTok, 153, 213
Title IX, xiii, xvi, 200–202, 204, 206, 216–17,
225
"To Understand Biological Sex, Look at the
Brain, Not the Body" (Boylan), 71
toxic masculinity, 87, 102, 146, 236
track and field, 3–4, 5, 25–26, 29–33, 109–10,
202, 238, 253
2009 World Championships in, 5, 6, 30–31
transgender, as term, periodic redefining of,
73–74, 75, 252
transgender community, 22, 25, 64, 68, 79,
101, 204, 209–10, 221, 222–23, 230, 240,
242, 244, 255
acceptance of sex as binary in, 66–67, 101–2,
214, 240
bathroom debate and, 18, 66, 71–72,
242–46
desire to procreate in, 97–98
discrimination against, xi, xii, 44, 66, 208–9
gender-affirming care for, *see* gender-affirming
care
normalizing of, 210
online harassment of, 223, 292n–93n
"passing" and, 257–58
respecting preferred pronouns of, 254–55
Transgender Journalists Association, 223–24
Style Guide of, 222–23, 227–28, 240
transgender kids, 25, 71, 203, 210, 215, 216,
240, 241
gender-affirming care for, xx, xxi, 18, 71, 216,
217, 218–19, 246–50
in school sports, 251–52
transgender men, 17, 44, 61, 66, 228, 231,
253, 256
on birth certificates, 46–47
pregnancy in, 26, 98, 228
transgender rights, xiv, xv, xviii, 44, 64–66, 78,
87, 196, 202, 203, 207, 209, 210, 211–19,
221, 240
in abortion debate, 65, 207

"cis" females as sidelined from discourse on,
222–24, 227
Equality Act and, xi, xii, xvii, 76–77, 199,
200, 201, 230
in feminist movement, xx–xxi, 64, 65, 66–67,
76–77, 203
landmark cases in, xxi, 44, 71–72, 209
political backlash to, xix–xx, 60, 216–19
private sector targeted in, 211–13
see also progressive advocacy
transgender women, 16, 18, 25–26, 30, 44, 65,
200, 213, 218, 230–31, 244, 246, 256
bathroom debate and, 244–46
definition of, 73–74, 256–57, 258–59
female sports and, xx, 60, 61–62, 69, 77, 79,
120, 201, 204, 205, 225, 245, 251–52,
253, 259–60
as lesbians, 101
military conscription and, 44–45
online harassment of, 223, 292n–93n
sexual orientation as "transphobic" claims
by, 100–102
"trans women are women" stance and, xviii,
60, 76, 79–80, 102, 203, 227, 260
Woman Question and, xviii, 60, 61–62, 64,
78–80, 256–61
Trans in America: Texas Strong (documentary), 25
transmedicalism, 259
"transphobic" label, 73, 112, 214, 222, 223–24,
225, 259
sexual orientation and, 100–102
Trevor Project, 218
"trigger" laws, 138
Trotter, William Monroe, 233
Trump, Donald, 203, 204, 215, 217–18
Truth, Sojourner, 60, 277n
Tryon, Milicent, 45
Tucker, Ross, 3
Turner, Brock, 145–46
Twitter (now X), 3, 64, 100, 213, 223, 225,
229–30, 292n–93n

Ukraine:
Russian invasion of, 136, 138–40,
162–63
surrogacy industry in, 140
undifferentiated gonadal ridge, 11–12
UN Human Rights Commission, 33
United Nations, 64, 243
United States v. Virginia (VMI), xix, xx, 46, 188,
189, 192–96, 198

universities, 219-21
U.S. News & World Report, 147
uterus, 12, 99, 231

vaccines, 132
Valieva, Kamila, 109
Vasari, Giorgio, 86, 107
Venus de Milo, 107-8, 241
Venus de Milo with Drawers (Dali), 107
Vice, 110
Victoria, Queen of England, 111
"Victory for Female Athletes Everywhere, A"
    (Coleman), 201-2
Villanova University, xiii, 201
Villa Romana del Casale (Sicily), 108
Vindication of the Rights of Woman, A
    (Wollstonecraft), 261-63, 264
Violence Against Women Act, xvi, 230
Virginia, 56
Virginia Military Institute (VMI), 192, 194
Vogue, British, 241
Volokh, Eugene, 225
voting rights, 170

Wall Street Journal, 132, 140, 151, 158
Walsh, Kyle, 129
Walsh, Matt, 223, 292n-93n
Walt Disney Company, 218
Washington, D.C., 147
Washington Post, 71, 77, 151, 152, 214, 224
    author's coauthored op-ed in, 200-201, 203,
    225
West Point, United States Military Academy
    at, 139
White, Byron, 180, 189
White, Sophie, 77-78
Winter, Jessica, 154, 155
Witch Trials of J. K. Rowling, The (podcast),
    226
Wollstonecraft, Mary, 261-64
Woman Question, 35, 53, 64, 68, 71, 78-80,
    176, 255-61
    in Jackson's confirmation hearings, 59-60,
    62-63, 77-78, 79

and redefining of "woman" and "female,"
    77-80, 256, 259, 260
transgender women and, xviii, 60, 61-62, 64,
    78-80, 256-61
Vindication and, 261-63, 264
women, *see* females
Women's Aid, 143
women's rights, xviii, xix, 35-36, 54, 64, 76,
    110, 169-70, 171, 173, 179, 180, 183,
    188, 196, 203, 211, 220-21, 263
    abortion and, *see* abortion, abortion rights
    in Afghanistan, 141
    Bradwell case and, 36-40, 42, 52, 53, 168
    voting and, 170
    see also feminism, feminists; sex
        classifications; sex discrimination
women's rights movement, 156
    black women as historically sidelined in, 60,
    173
    in nineteenth and early twentieth century,
    169-70, 262
    sex equality as not initially goal of, 169-70
    see also feminism, feminists
Woo, Ilyon, 169, 257
Woodstock, N.Y., 234
workplace/employment, xxi, 141, 156, 170
    males' declining participation in, 158-61
    onsite childcare in, 150-51
    sex discrimination in, 34-40, 44, 52, 54,
    171-72, 183
World Anti-Doping Agency (WADA), xiv
World Aquatics (formerly FINA (Fédération
    Internationale de Natation), 114-15
World Athletics (formerly International
    Amateur Athletic Federation), 30
World Health Organization (WHO), 8, 9, 22,
    145

Yale University, 35, 245
Yellen, Janet, 137
youth and community sports, 115, 119, 121,
    250-52

Zhou, Viola, 110

About the Author

DORIANE LAMBELET COLEMAN is professor at Duke Law School, where she specializes in interdisciplinary scholarship focused on women, sports, children, and law. Her work has been published in numerous journals, and she is regularly cited in the press. At Duke, she is a faculty fellow and member of the advisory council of the Kenan Institute for Ethics; a faculty associate of the Trent Center for Bioethics, Humanities & History of Medicine at the School of Medicine; a member of the Athletic Council; and co-director of the law school's Center for Sports Law and Policy. She received a juris doctor from Georgetown University Law Center in 1988 and a bachelor of arts degree from Cornell University in 1982. Before attending law school, Coleman ran the 800 meters for Cornell and Villanova, and as a professional for Athletics West, the Santa Monica Track Club, the Atoms Track Club, and the Swiss and U.S. national teams.